T0231845

MATERIALS PHYSICS AND CHEMISTRY

Applied Mathematics and Chemo-Mechanical Analysis

MATERIALS PHYSICS AND CHEMISTRY

Applied Mathematics and Chemo-Mechanical Analysis

Edited by

Satya Bir Singh, PhD
Alexander V. Vakhrushev, DSc
A. K. Haghi, PhD

Apple Academic Press Inc.
4164 Lakeshore Road
Burlington ON L7L 1A4
Canada

Apple Academic Press Inc.
1265 Goldenrod Circle NE
Palm Bay, Florida 32905
USA

First issued in paperback 2021

Library and Archives Canada Cataloguing in Publication

Title: Materials physics and chemistry : applied mathematics and chemo-mechanical analysis / edited by Satya Bir Singh, PhD, Alexander V. Vakhrushev, DSc, A.K. Haghi, PhD.

Names: Singh, Satya Bir, editor. | Vakhrushev, Alexander V., editor. | Haghi, A. K., editor.

Description: Includes bibliographical references and index.

Identifiers: Canadiana (print) 2020027029X | Canadiana (ebook) 20200270486 | ISBN 9781771888677 (hardcover) | ISBN 9780367816094 (ebook)

Subjects: LCSH: Solids—Mathematical models. | LCSH: Solids—Properties. | LCSH: Solid state physics. | LCSH: Strength of materials. | LCSH: Mechanics, Applied.

Classification: LCC QC176 .M37 2020 | DDC 530.4/1—dc23

Library of Congress Cataloging-in-Publication Data

Names: Singh, Satya Bir, editor. | Vakhrushev, Alexander V., editor. | Haghi, A. K., editor.

Title: Materials physics and chemistry : applied mathematics and chemo-mechanical analysis / edited by Satya Bir Singh, PhD, Alexander V. Vakhrushev, DSc, A.K. Haghi, PhD.

Description: 1st edition. | Burlington ON, Canada ; Palm Bay, Florida : Apple Academic Press, [2021] | Includes bibliographical references and index. | Summary: "This volume focuses on the development and application of fundamental concepts in mechanics and physics of solids as they pertain to the solution of challenging new problems in diverse areas, such as materials science and micro- and nanotechnology. In this volume, emphasis is placed on the development of fundamental concepts of mechanics and novel applications of these concepts based on theoretical, experimental, or computational approaches, drawing upon the various branches of engineering science and the allied areas within applied mathematics, materials science, and applied physics. Introducing and demonstrating the utility of computational tools and simulations, this volume provides a modern view of how solids behave, how they can be experimentally characterized, and how to predict their behavior in different load environments. It discusses recent developments and compares them to classical theories along with introducing readers to the current state of the art in predicting failure using a combination of experiment and computational techniques. Materials Physics and Chemistry: Applied Mathematics and Chemo-Mechanical Analysis emphasizes the basics, such as design, equilibrium, material behavior, and geometry of deformation in simple structures or machines. Readers will find a thorough treatment of stress, strain, and the stress-strain relationships. Meanwhile it provides a solid foundation upon which readers can begin work in composite science and engineering. Many chapters include theory components with the equations students need to calculate different properties. Key features: Emphasizes using math to solve engineering problems instead of focusing on derivations and theory Encompasses the full range of applied mathematics and engineering from modeling to solution, both analytical and numerical. Develops a framework for the equations and numerical methods of applied mathematics"-- Provided by publisher.

Identifiers: LCCN 2020025630 (print) | LCCN 2020025631 (ebook) | ISBN 9781771888677 (hardcover) | ISBN 9780367816094 (ebook)

Subjects: LCSH: Materials science--Mathematical models.

Classification: LCC TA403.6 .M3735 2021 (print) | LCC TA403.6 (ebook) | DDC 620.1/12--dc23

LC record available at https://lccn.loc.gov/2020025630

LC ebook record available at https://lccn.loc.gov/2020025631

Apple Academic Press also publishes its books in a variety of electronic formats. Some content that appears in print may not be available in electronic format. For information about Apple Academic Press products, visit our website at **www.appleacademicpress. com** and the CRC Press website at **www.crcpress.com**

About the Editors

Satya Bir Singh, PhD

Professor, Department of Mathematics, Punjabi University, Patiala, India

Satya Bir Singh, PhD, is a Professor of Mathematics at Punjabi University Patiala in India. Prior to this, he has worked as an assistant professor in Mathematics at the Thapar Institute of Engineering and Technology, Patiala, India. He has published about 125 research papers in journals of national and international repute and has given invited talks at various conferences and workshops and has also organized several national and international conferences. He has been a coordinator and principal investigator of several schemes funded by the Department of Science and Technology, Government of India, New Delhi; the University Grants Commission, Government of India, New Delhi; and the All India Council for Technical Education, Government of India, New Delhi. He has 21 years of teaching and research experience. His areas of interest include mechanics of composite materials, optimization techniques, and numerical analysis. He is a life member of various learned bodies.

Alexander V. Vakhrushev, DSc

Professor, M. T. Kalashnikov Izhevsk State Technical University, Izhevsk, Russia; Head, Department of Nanotechnology and Microsystems of Kalashnikov Izhevsk State Technical University, Russia

Alexander V. Vakhrushev, DSc, is a Professor at the M. T. Kalashnikov Izhevsk State Technical University in Izhevsk, Russia, where he teaches theory, calculating, and design of nano- and microsystems. He is also the chief researcher of the Department of Information-Measuring Systems of the Institute of Mechanics of the Ural Branch of the Russian Academy of Sciences and the Head of the Department of Nanotechnology and Microsystems of Kalashnikov Izhevsk State Technical University. He is a corresponding member of the Russian Engineering Academy. He has over 400 publications to his name, including monographs, articles, reports, reviews, and patents. He has received several awards, including an Academician A. F. Sidorov Prize from the Ural Division of the Russian

Academy of Sciences for significant contribution to the creation of the theoretical fundamentals of physical processes taking place in multilevel nanosystems and was proclaimed an honorable scientist of the Udmurt Republic. He is currently a member of editorial boards of several journals, including *Computational Continuum Mechanics, Chemical Physics and Mesoscopia*, and *Nanobuild*. His research interests include multiscale mathematical modeling of physical–chemical processes into the nano-hetero systems at nano-, micro- and macrolevels; static and dynamic inter-action of nanoelements; and basic laws relating the structure and macro characteristics of nano-hetero structures.

A. K. Haghi, PhD

Professor Emeritus, Engineering Sciences, Former Editor-in-Chief, International Journal of Chemoinformatics and Chemical Engineering and Polymers Research Journal; Member, Canadian Research and Development Center of Sciences and Cultures

A. K. Haghi, PhD, is the author and editor of 165 books, as well as 1000 published papers in various journals and conference proceedings. Dr. Haghi has received several grants, consulted for a number of major corporations, and is a frequent speaker to national and international audiences. Since 1983, he served as a professor at several universities. He is former Editor-in-Chief of the *International Journal of Chemoinformatics and Chemical Engineering and Polymers Research Journal* and is on the editorial boards of many international journals. He is also a member of the Canadian Research and Development Center of Sciences and Cultures (CRDCSC), Montreal, Quebec, Canada. He holds a BSc. in urban and environmental engineering from the University of North Carolina (USA), an MSc in mechanical engineering from North Carolina A&T State University (USA), a DEA in applied mechanics, acoustics and materials from the Université de Technologie de Compiègne (France), and a PhD in engineering sciences from Université de Franche-Comté (France).

Contents

Contributors

Zainab Al Ani
Department of Petroleum & Chemical Engineering, College of Engineering,
Sultan Qaboos University, PO Box 33, Al-Khod, P.C. 123, Muscat, Sultanate of Oman

Talal Al Wahaibi
Department of Petroleum & Chemical Engineering, College of Engineering,
Sultan Qaboos University, PO Box 33, Al-Khod, P.C. 123, Muscat, Sultanate of Oman

P. Benrajesh
Department of Mechanical and Automobile Engineering, CHRIST (Deemed-to-be-University),
Kanminike, Kumbalgodu, Mysore Road, Kengeri, Bangalore 560074, Karnataka, India

Tania Bose
Department of Mathematics, UIET, Panjab University, Chandigarh, India

Tanmoy Chakraborty
Department of Chemistry, Manipal University Jaipur 303007, India.
E-mail: tanmoychem@gmail.com, tanmoy.chakraborty@jaipur.manipal.edu

Neeraj Chamoli
Department of Mathematics, PG DAV College, Chandigarh, India

A. Y. Fedotov
Department "Mechanics of Nanostructures," Institute of Mechanics, Udmurt Federal Research
Center, Ural Division, Russian Academy of Sciences, Izhevsk, Russia
Department "Nanotechnology and Microsystems," Technic Kalashnikov Izhevsk State Technical
University, Izhevsk, Russia

Ashish M. Gujarathi
Department of Petroleum and Chemical Engineering, Sultan Qaboos University, Muscat 123, Oman.
E-mail: ashishg@squ.edu.om

Chandan Guria
Department of Petroleum Engineering, Indian Institute of Technology (Indian School of Mines),
Dhanbad 826004, India

G. S. Hebbar
Department of Mechanical Engineering, Faculty of Engineering, CHRIST
(Deemed to be University), Bangalore 560074, India

Chaudhery Mustansar Hussain
Department of Chemistry and Environmental Sciences, New Jersey Institute of Technology,
University Heights, Newark, NJ 07102, USA. E-mail: chaudhery.m.hussain@njit.edu

K. Shiva Kumar
Department of Mechanical and Automobile Engineering, CHRIST (Deemed-to-be-University),
Kanminike, Kumbalgodu, Mysore Road, Kengeri, Bangalore 560074, Karnataka, India

P. Pal Pandian
Department of Mechanical Engineering, Faculty of Engineering, CHRIST
(Deemed to be University), Bangalore 560074, India

Sukanchan Palit
Department of Chemical Engineering, University of Petroleum and Energy Studies,
Energy Acres, Dehradun 248007, Uttarakhand, India

Prabhat Ranjan
Department of Mechatronics Engineering, Manipal University Jaipur 303007, India

Minto Rattan
Department of Mathematics, UIET, Panjab University, Chandigarh, India

G. Ravichandran
Department of Mechanical Engineering, Faculty of Engineering, CHRIST
(Deemed to be University), Bangalore 560074, India. E-mail: ravig_s@rediffmail.com

Ivan Sunit Rout
Department of Mechanical Engineering, Faculty of Engineering, CHRIST
(Deemed to be University), Bangalore 560074, India. E-mail: ivan.rout@christuniversity.in

M. Sadashiva
Department of Mechanical Engineering, P.E.S.C.E., Karnataka, India.
E-mail: sadashiva015@gmail.com

N. Santhosh
Department of Mechanical and Automobile Engineering, Christ (Deemed to be University),
Bangalore, Karnataka, India

Monika Sethi
Department of Mathematics, Faculty of Science and Technology, ICFAI University,
Baddi Solan 174103, Himachal Pradesh, India

Shivdev Shahi
Department of Mathematics, Punjabi University Patiala, Punjab, India.
E-mail: shivdevshahi93@gmail.com

V. Sharanraj
Department of Mechanical Engineering, S. J. Polytechnic, Bangalore, Karnataka, India

A. V. Shushkov
Department "Mechanics of Nanostructures," Institute of Mechanics,
Udmurt Federal Research Center, Ural Division, Russian Academy of Sciences,
Izhevsk, Russia; Department "Nanotechnology and Microsystems,"
Technic Kalashnikov Izhevsk State Technical University, Izhevsk, Russia

Satya Bir Singh
Department of Mathematics, Punjabi University, Patiala, Punjab, India.
E-mail: sbsingh69@yahoo.com

M. R. Srinivasa
Department of Mechanical Engineering, P.E.S.C.E., Karnataka, India

A. Temesgen
Department of Mathematics, Punjabi University, Patiala, Punjab, India

Pankaj Thakur
Department of Mathematics, Faculty of Science and Technology, ICFAI University,
Baddi Solan 174103, Himachal Pradesh, India. E-mail: pankaj_thakur15@yahoo.co.in

Debasish Tikadar
Department of Petroleum and Chemical Engineering, Sultan Qaboos University,
Muscat 123, Oman; Department of Petroleum Engineering, Indian Institute of Technology
(Indian School of Mines), Dhanbad 826004, India; Worley Oman Engineering LLC,
Muscat 133, Oman

Chefi Triki
Department of Mechanical and Industrial Engineering, College of Engineering,
Sultan Qaboos University, PO Box 33, Al-Khod, P.C. 123, Muscat, Sultanate of Oman

A. V. Vakhrushev
Department "Mechanics of Nanostructures," Institute of Mechanics,
Udmurt Federal Research Center, Ural Division, Russian Academy of Sciences,
Izhevsk, Russia; Department "Nanotechnology and Microsystems,"
Technic Kalashnikov Izhevsk State Technical University, Izhevsk, Russia.
E-mail: vakhrushev-a@yandex.ru

G. Reza Vakili-Nezhaad
Department of Petroleum & Chemical Engineering, College of Engineering,
Sultan Qaboos University, PO Box 33, Al-Khod, P.C. 123, Muscat, Sultanate of Oman

L. Francis Xavier
Assistant Professor, Department of Mechanical and Automobile Engineering,
CHRIST (Deemed-to-be-University), Kanminike, Kumbalgodu, Mysore Road, Kengeri,
Bangalore 560074, Karnataka, India. E-mail: francis.xavier@chirstuniversity.in

A. V. Zemskov
Department "Nanotechnology and Microsystems," Technic Kalashnikov Izhevsk State Technical
University, Izhevsk, Russia

Abbreviations

AP	acidification potential
CDFT	conceptual density functional theory
CMC	ceramic matrix composites
CSA	coconut shell ash
DEA	diethanolamine
DFT	density functional theory
EDM	electric discharge machining
EMOO	excel-based multiobjective optimization
FGM	functionally graded material
FSW	friction stir welding
GWP	global warming potential
HAP	hydroxyapatite
HOMO	highest occupied molecular orbital
I2SI	integrated inherent safety index
I-MODE	multi-objective differential evaluation
LRT	linear reciprocating tribometer
LSDA	local spin density approximation
LUMO	lowest unoccupied molecular orbital
MDEA	methyldiethanolamine
MEA	monoethanolamine
MMC	metal matrix composite
MOO	multi-objective optimization
NSGA	nondominated sorting genetic algorithm
PMC	polymer matrix composite
PRS	maximization of product sales
SEM	scanning electron microscope
SLS	selective-laser sintering
TPC	total production cost

Preface

Mechanics has had a rich history in defining, inspiring, and enabling innovation in many scientific and technological fields. Because it provides a common rigorous foundation for numerous science and engineering disciplines, its imprint can be found in numerous critical bodies of knowledge that attempt to understand, predict, and affect the world around us. This volume focuses on the development and application of fundamental concepts in mechanics and physics of solids as they pertain to the solution of challenging new problems in diverse areas such as materials science, micro, and nanotechnology. In this volume, emphasis is placed on the development of fundamental concepts of mechanics and novel applications of these concepts based on theoretical, experimental, or computational approaches, drawing upon the various branches of engineering science and the allied areas within applied mathematics, materials science, and applied physics.

Elasticity, plasticity, damage mechanics, and cracking are all phenomena that determine the resistance of solids to deformation and fracture. Introducing and demonstrating the utility of computational tools and simulations, this volume provides a modern view of how solids behave, how they can be experimentally characterized, and how to predict their behavior in different load environments. It will discuss recent developments and compare them to classical theories along with introducing readers to the current state of the art in predicting failure using a combination of experiment and computational techniques.

This volume emphasizes the basics, such as design, equilibrium, material behavior, and geometry of deformation in simple structures or machines. Readers will find a thorough treatment of stress, strain, and stress–strain relationships.

Meanwhile, it provides a solid foundation upon which readers can begin work in composite materials science and engineering.

Many chapters include theory components with the equations students need to calculate different properties.

In the first chapter, classical and nonclassical treatment of problems in elastic-plastic and creep deformation for rotating discs discussed in detail. In this chapter, exhaustive literature review has been reviewed on the analysis of elastic-plastic and creep deformation in rotating disc using classical and nonclassical treatment. In classical treatment, researchers used simplifying empirical assumptions like yield criterion and the associated flow rule for elastic-plastic transition and creep strain laws (or a power relationship between stress and strain) for creep transition. However, in nonclassical treatment, researchers used the transition theory combined with the generalized strain measure, which does not require any yield condition or creep law that means they used Seth's transition and generalized strain measure theory.

Chapter 2 formulates a model describing the effect of thermal gradation on the steady-state creep in a rotating anisotropic functionally graded disc of aluminum having linearly varying SiC_p from inner to outer radius. Modeling has been done for discs operating at parabolic temperature profile and the results are compared with disc acting at uniform temperature. Consequently, the study of thermal gradation in anisotropic functionally graded rotating discs has been made to give new insights to material engineers and designers for applications in gas turbines, jet engines, and other dynamic operators.

On the basis of numerical results obtained in Chapter 3, it has been shown that enamel and dentine under uniaxial compression behaves like a functionally graded strong hard tissue with necessary elastic and plastic limit. It demonstrates a considerable ability to suppress a crack growth. Different values of pressure required for initial yielding and fully plastic state were calculated for various radius ratios depending on the geometry of the sample. Trends of the graphs were similar for enamel and hydroxyapatite due to enamel's composition. Significant difference between stress buildup at crown and root dentine has been observed by varying the radii ratios in the modeled spherical shell. The mathematical model developed to analyze the transversely isotropic behavior of enamel and dentine may be of great use in manufacture of highly competitive dental implants.

Characterization of material in solid disc is presented in Chapter 4 in detail.

In Chapter 5, characterization of material in a rotating disc subjected to thermal gradient is reviewed and investigated.

Optimization of cutting parameters for hard machining on Inconel 718 using signal to noise ratio and gray relational analysis is presented in Chapter 6.

Chapter 7 is devoted to experimental investigations of the elastic modulus and hardness dependences of tungsten carbide deposited by laser high-speed sintering on stainless steel substrate and surface topology. The researches of mechanical properties were fulfilled by the indentation method on studying complex system of physical and mechanical properties of materials Nanotest 600. Surface topology on the noncontact optical profilometer New View 6300 was investigated. Hardness, Young's modulus increases in the passage trajectory of the laser because of the quenching phenomenon. The purpose of this chapter is an experimental study of the mechanical characteristics and roughness of a sample with a powder coating based on tungsten carbide deposited on a stainless steel substrate and treated with a laser beam.

In Chapter 8, we have reported the physical and chemical properties of gold–vanadium nanoalloy clusters invoking density functional theory methodology. The density functional-based descriptors, namely, highest occupied molecular orbital (HOMO)–lowest unoccupied molecular orbital (LUMO) energy gap, molecular hardness, softness, electronegativity, and electrophilicity index have been computed. The computed data reveals that HOMO–LUMO energy gap of the Au–V clusters have direct relationship with molecular hardness values and inverse relation with molecular softness data. The nanoalloy clusters having large HOMO–LUMO energy gap is more stable and less reactive. This is an expected trend from experimental point of view, as hardness of cluster increases with increase in frontier orbital energy gap. The high value of linear regression coefficient between electrophilicity index and HOMO–LUMO energy gap supports our computational study.

The aim of Chapter 9 is to develop an algorithm for calculating the uniformity of mixing of a two-component mixture of micro- and nanoelements and the creation of a software and hardware complex that allows for operative control of the mixing process.

The simultaneous multicriteria based optimization trend in industrial cases is discussed in Chapter 10 in detail.

Chapter 11 is devoted to simultaneous optimization aspects in industrial gas sweetening process for sustainable development.

In Chapter 12, a theoretical study has been performed on Fe-doped Cu nanoalloy clusters Cu_nFe ($n = 1$–5) by using Density Functional Theory (DFT) methodology. The DFT-based global descriptors, namely, molecular hardness, softness, electronegativity, electrophilicity index, and dipole moment have been calculated along with their HOMO–LUMO energy gap. The result indicates that cluster Cu_3Fe has maximum HOMO–LUMO energy gap, whereas cluster Cu_5Fe shows minimum energy gap in this range. The computed result implies that HOMO–LUMO energy gap runs hand in hand along with the molecular hardness of Cu_nFe clusters. This is a probable propensity from experimental point of view also, as the frontier orbital energy gap increases, their molecular hardness value also increases. The computed regression coefficient between DFT-based descriptors and HOMO–LUMO energy gap supports this study.

Nanotechnology as a clean technology and a vision for the future is presented in Chapter 13.

In Chapter 14, Aluminum 6063 alloy was selected as the base matrix material in preparing the composite material. Aluminum 6063 alloy is used in various applications such as heat sink sections, flexible assembly systems, special machinery elements, railings, truck and trailer flooring, radiator, and other heat-exchanger applications.

Objectives of Chapter 15 are as follows:

1. Selection of proper matrix material and reinforcements to produce high-standard metal-matrix composite having superior properties and behavior.
2. To fabricate high-quality metal-matrix composite material through stir casting.
3. To weld the composite plates so as to obtain sound weld joint using friction stir welding.
4. To create standard fretting wear test specimens conforming to ASTM standards using wire electric discharge machining.
5. To conduct fretting wear test using linear reciprocating tribometer.

This book serves as a reference for postgraduate students and as a reference for practitioners using linear and nonlinear analysis in engineering and design.

Features

- Each chapter emphasizes using math to solve engineering problems instead of focusing on derivations and theory.
- Encompasses the full range of applied mathematics and engineering from modeling to solution, both analytical and numerical.
- It develops a framework for the equations and numerical methods of applied mathematics.

CHAPTER 1

Classical and Nonclassical Treatment of Problems in Elastic-Plastic and Creep Deformation for Rotating Discs

A. TEMESGEN[1], S. B. SINGH[2*], and PANKAJ THAKUR[3]

1Department of Mathematics, Punjabi University, Patiala, Punjab, India

2Department of Mathematics, Punjabi University, Patiala, Punjab, India

3Department of Mathematics, ICFAI University, Kallujhanda, Himachal Pradesh, India

*Corresponding author. E-mail: sbsingh69@yahoo.com

ABSTRACT

Rotating discs are found in a range of engineering applications such as steam turbines, gas turbine engines, flywheels, automobile engine, turbo generator, gears, compressors, disc brakes of cars, shrink fits, circular saws, storage devices (hard disks, blue ray disks, etc.), computer disc drivers, and so on. The use of rotating discs in engineering application has generated considerable interest in many problems and has made a longstanding research area in the domain of mechanics of solid so that researchers devoted to the analysis of stresses in rotating discs to achieve the optimum design of structural components.

In this chapter, exhaustive literature review has been reviewed on the analysis of elastic-plastic and creep deformation in rotating disc using classical and nonclassical treatment. In classical treatment, researchers used simplifying empirical assumptions like yield criterion and the associated flow rule for elastic-plastic transition and creep strain laws (or a power relationship between stress and strain) for creep transition. However, in nonclassical treatment, researchers used the transition theory combined with the generalized strain measure which does not require any yield condition or creep law that means they used Seth's transition and generalized strain measure theory. The nonclassical treatment of an elastic-plastic and creep transition in rotating disc was started after 1962 and after 1972, respectively.

The review of the literature reveals that the analysis of stresses and strains in rotating disc has been extensively performed with regard to elastic-plastic and creep transition in classical treatment having constant thickness and especially variable thickness in an isotropic and transversely isotropic rotating disc. However, the studies using the nonclassical treatment of rotating discs are rather limited as compared with classical treatment even though the real behavior of the transition of the material is nonlinear and the method neglects the empirical assumptions.

Recent studies from the literature review indicate that elastic-plastic transitional stress distribution and displacement for transversely isotropic rotating discs subject to mechanical load having variable thickness or variable density, creep stresses and strains for transversely isotropic rotating disc having constant thickness, and creep stresses and strains for a thin-rotating disc having variable thickness and variable density with edge load have been studied.

In view of this, there is a need to investigate elastic-plastic and creep transitional stresses and strains for transversely isotropic/orthotropic rotating disc having variable thickness or variable density under different loading such as thermal gradient, internal pressure, and external pressure.

1.1 INTRODUCTION

Mechanics is the science of force and motion of matter. Solid mechanics is the science of force and motion of matter in the solid-state. Solid mechanics is the oldest field of science and is still advancing rapidly in all areas of technology. The advancement of the methods of analysis of structures of various kinds over the years is indeed quite impressive. Solid mechanics has a great role on engineering structures and machines, such as airplanes, automobiles, bridges, spacecraft, buildings, electric generators, gas turbines, and so forth, are usually formed by connecting various parts or members. In most structures or machines, the primary function of a member is to support or transfer external forces (loads) that act on it, without failing. Failure of a member may occur when it is loaded beyond its capacity to resist fracture, general yielding, excessive deflection, or instability. These types of failure depend on the nature of the load and the type of member.

Solid mechanics is a diverse science with roots in traditional research fields, while also being an integral part of many entirely new areas. Much work is going on in classical research areas such as structural mechanics, where also many PhDs in solid mechanics use their expertise to solve complicated mechanics problems in private companies. This includes offshore structures, large ships, modern bridges, and very tall buildings. Related to this is the analysis of large shell structures, and there is also a big demand for expertise on contact and friction mechanics, which are important research areas with many unresolved issues.

In the current research, solid mechanics is a fundamental discipline which addresses as yet poorly understood phenomenon in the mechanical response and failure of materials and structures. Research in solid mechanics is essential not only for basic understanding of mechanical phenomena but also to advance engineering methodology in a host of areas throughout mechanical and structural technology. Advances in the subject are central to assuring safety, reliability, and economy in design of structures, devices, machines, and complete systems, and hence to the continued development of power generation technologies such as fusion, nuclear and gas turbine power, aerospace and surface transportation vehicles, earthquake-resistant design, offshore structures, orthopedic devices, and materials processing and manufacturing technologies. Nowadays, hot research areas in solid mechanics are material mechanics, computational mechanics, dynamics, instabilities in structures, biomechanics, manufacturing, optimal design of engineering structural components, and so on.

In all areas of solid mechanics research, whether related to advanced engineering methodology or fundamental phenomena, the impact of modern computational mechanics has been enormous. Nearly all the fundamental and applied research areas will profit from continued adoption of computational techniques and widespread availability of advanced computing systems to solid mechanics researchers for the vast range of complex and nonlinear mechanical response that is not yet routinely analyzable. Most of the physical concepts used in solid mechanics are modeled by mathematical entities known as tensors.

1.1.1 *MECHANICS OF DEFORMABLE BODIES*

An application of fundamental law of mechanics to real-life problem is known as engineering mechanics. The engineering mechanics is subdivided into two classes: (1) solid mechanics, (2) fluid mechanics, based on the object on which the mechanics is imposed. Further, the solid mechanics is subdivided into two major branches, that is, mechanics of rigid bodies and mechanics of deformable bodies. The body in which deformations are negligible is known as rigid body. The mechanics of *undeformed (rigid)* bodies, which deals with the body in state of rest, is known as statics and the body in the state of motion is known as dynamics. Again, the dynamic state is categorized into two parts one is kinematics which describes the motion of the object with any cause and second one is kinetics which deals with motion of the object or body under the action of force.

In this study, we focus on mechanics of deformable bodies. This field further classified as theory of elasticity, theory of plasticity, and theory of creep. Solid mechanics, which includes the theories of elasticity, plasticity, and creep, is a broad discipline, with experimental, theoretical, and computational aspects, and with a twofold aim:

1. It seeks to describe the mechanical behavior of solids under conditions as general as possible, regardless of shape, interaction with other bodies, field of application, or the like.
2. It attempts to provide solutions to specific problems involving stressed solid bodies that arise in civil and mechanical engineering, geophysics, physiology, and other applied disciplines.

1.1.1.1 *THE CONTINUOUS MEDIUM*

The mechanics of continuous medium is that branch of mechanics concerned with the stresses in solids, liquids, and gases and the deformation or flow of these materials. The adjective continuous refers to the simplifying concept underlying the analysis: we disregard the molecular structure of matter and picture it as being without gaps or empty spaces. We further suppose that all the mathematical functions entering the theory are continuous functions, except possibly at a finite number of interior surfaces separating regions of continuity. This statement implies that the derivatives of the functions are continuous too, if they enter the theory, since all functions entering the theory are assumed continuous. This hypothetical continuous material is called as continuous medium or continuum. Continuous medium permits us to define stress at a point. A material is continuous if it completely fills the space that it occupies, leaving no pores or empty space, and further its properties are describable by continuous functions. A homogeneous material has identical properties at all points.

1.1.1.2 METHODS OF ANALYSIS

For a given member subjected to prescribed loads, the load–stress relations are based on the following requirements:

a) The equations of equilibrium (or equations of motion for bodies not in equilibrium).
b) The compatibility conditions (continuity conditions) that require deformed volume.
c) The constitutive relations elements in the member to fit together without overlap or tearing.

Two different methods are used to satisfy requirements: (a) the method of mechanics of materials and (b) the method of general continuum mechanics.

1.1.1.2.1 Method of Mechanics of Materials

A simple member such as a circular shaft of uniform cross-section may be subjected to complex loads that produce a *multiaxial* state of stress. However, such complex loads can be reduced to several simple types of load, such as axial, bending, and torsion. Each type of load, when acting alone, produces mainly one stress component, which is distributed over the cross-section of the member. The method of mechanics of materials can be used to obtain load–stress relations for each type of load. If the deformations of the member that result from one type of load do not influence the magnitudes of the other types of loads and if the material remains linearly elastic for the combined loads, the stress components resulting from each type of load can be added together (i.e., the method of superposition may be used).

The method of mechanics of materials is based on simplified assumptions related to the geometry of deformation (requirement (b)) so that strain distributions for a cross-section of the member can be determined. A basic assumption is that plane sections before loading remains plane after loading. In a similar way, we often assume that lines normal to the middle surface of an underformed plate remain straight and normal to the middle surface after the load is applied.

In a complex member, each load may have a significant influence on each component of the state of stress. Then, the method of mechanics of materials becomes cumbersome and the use of the method of continuum mechanics may be more appropriate.

1.1.1.2.2 Method of Continuum Mechanics

Many of the problems treated in solid mechanics have multiaxial states of stress of such complexity that the mechanics of materials method cannot be employed to derive load–stress and load–deflection relations. Therefore, in such cases, the method of continuum mechanics is used. When we consider small displacements and linear elastic material behavior only, the general method of continuum mechanics reduces to the method of the theory of linear elasticity. In the derivation of load–stress and load–deflection relations by the theory of linear elasticity, an infinitesimal volume element at a point in a body with faces normal to the coordinate axes is often employed. Requirement (a) is represented by the differential equations of equilibrium. Requirement (b) is represented by the differential equations of compatibility. The material response (requirement (c)) for linearly elastic behavior is determined by one or more experimental tests that define the required elastic coefficients for the material. These coefficients can be obtained from a tension specimen if both axial and lateral strains are measured for every load applied to the specimen. Requirement (c) is represented therefore by the isotropic or transversely isotropic or anisotropic stress–strain relations developed. If the differential equations of equilibrium and the differential equations of compatibility can be solved subject to specified stress–strain relations and specified boundary conditions, the states of stress and displacements for every point in the member are obtained.[6,39]

1.1.1.3 BASIC CONCEPTS OF ELASTIC, PLASTIC, AND CREEP

Consider an elastic solid which is subjected to some small external loads for which the deformation is elastic, that is, upon the release of these loads the body resumes its initial unstressed and undeformed state. The range of stresses and strains for which this is true is known as the elastic range.

Elasticity is a subject that deals with determination of the stress, strain, and displacement distribution in an elastic solid under the influence of external forces. Elasticity theory establishes a mathematical model of the deformation problem, and this requires mathematical knowledge to understand the formulation and solution procedures. Governing partial differential field equations are developed using basic principles of continuum mechanics commonly formulated in vector and tensor. Techniques used to solve these field equations can encompass Fourier methods, *variational* calculus, integral transforms, complex variables, potential theory, finite differences, finite elements, and so on.

Experiments indicate that when the external loads are increased gradually, beyond a certain value of loads, the elastic character of the body is destroyed, that is, upon the release of loads the body will not resumes its original state and so retains some permanent deformations. This permanent deformation is known as plastic deformation (irreversible permanent molecular rearrangement).

Similarities between elasticity and plasticity:

- Both are properties of solid.
- Any type of loading (normal, shear, or mixed) may result both types of deformations.
- Plastic deformation can occur only after the material is elastically deformed. So without elastic deformation, plastic deformation is not possible.
- Both elastic and plastic deformations are useful; however, based on the application.

Differences between elasticity and plasticity are given in Table 1.1.

Most of solid materials show noticeable elastic and plastic deformations. These materials are usually named as elastoplastic (or elastic-plastic) materials. Analysis of structural components on the basis of elastoplastic diagram is called elastoplastic analysis or plastic analysis.

If a bar is subjected to a uniform tension or compression, that is, a direct force, which is uniformly or equally applied across the cross-section, then the internal forces set up are also distributed uniformly and the bar is said to be subjected to a uniform direct or normal stress, the stress being defined as

TABLE 1.1 Differences between Elasticity and Plasticity.

Elasticity	Plasticity
Elasticity is the property of the solid material by virtue of which it tends to regain its shape after the removal of external load	Plasticity is the property of the solid material by virtue of which it tends to retain its deformed shape even after the removal of external load
In elastic deformation, although atoms of the solid are displaced from their original lattice site, they return back to their original position once external loading is removed. So atoms are temporarily displaced	In plastic deformation, atoms of the solid are permanently displaced from their original lattice site. They retain the new position even after the removal of external loading
External force required for elastic deformation of solid is quite small	Force required for plastic deformation is also higher
Hooke's law of elasticity is applicable within this elastic region	Hooke's Law is not applicable if the material is plastically deformed
Linear stress–strain behavior within elastic region	Stress–strain curve is nonlinear in plastic region
Energy absorbed by the material during elastic deformation is called module of resilience, i.e., high yield strength	Total energy absorbed by the material during elastic and plastic deformation region is called module of toughness, i.e., don't have the ability to absorb energy up to a fracture
Mechanical and metallurgical properties of the solid material remain unaltered when it is elastically deformed	Many properties of the solid material change considerably for plastic deformation
Elasticity is an important consideration for machine tool structures, bridges, other civil frames, equipment body, many household structures and frames that should retain their shape	Plasticity is an important consideration for sheet metal working, various forming operations, such as rolling, forging, extrusion, etc., rivet joining

$$\text{Stress}\left(\sigma\right) = \frac{\text{Load}}{\text{Area}} = \frac{P}{A}$$

Stress σ may thus be compressive or tensile depending on the nature of the load and will be measured in units of Newton's per square meter or multiples of this. In some cases, the loading situation is such that the stress will vary across any given section, and in such cases, the stress at any point is given by the limiting value of $\delta P/\delta A$ as δA tends to zero.

If a bar is subjected to a direct load, and hence a stress, the bar will change in length. If the bar has an original length L and changes in length by an amount δL, the strain produced is defined as follows:

$$\text{Strain}\left(\varepsilon\right) = \frac{\text{Change in length}}{\text{Original length}} = \frac{\delta L}{L}$$

Strain is thus a measure of the deformation of the material.

If materials are elastic, stress is proportional to strain, Hooke's law.

$$\frac{\sigma}{\varepsilon} = E$$

where σ is a stress, ε is a strain, and E is modulus of elasticity or Young's modulus. E is generally assumed to be the same in tension or compression and for most engineering materials has a high numerical value. The actual value of Young's modulus for any material is normally determined by carrying out a standard tensile test on a specimen of the material as described in Figure 1.1.

1.1.1.3.1   *Tensile Test*

To compare the strengths of various materials, it is necessary to carry out some standard form of test to establish their relative properties. One such test is the standard tensile test in which a circular bar of uniform cross-section is subjected to a gradually increasing tensile load until failure occurs. Curve 1, obtained from tensile test in Figure 1.1, illustrates nominal stress–strain diagram in which the stresses are usually computed on the basis of the original area of the specimen; such stresses are often referred to as conventional or nominal stresses whereas Curve 2 illustrates true stress–strain diagram. Since when a material is subjected to a uniaxial load, some contraction or expansion always takes place. Thus, dividing the applied force by the corresponding actual area of the specimen at the same instant gives the so-called true stress.

Measurements of the change in length of a selected gauge length of the bar are recorded throughout the loading operation by means of extensometers and a graph of load against extension or stress against strain is produced as shown in Figure 1.1; this shows a typical result for a test on a mild (low carbon) steel bar; other materials will exhibit different graphs but of a similar general form. According to Curve 1, for the first part of the test, it will be observed that Hooke's law is obeyed, that is, the material behaves elastically and stress is proportional to strain, giving the straight line graph indicated.

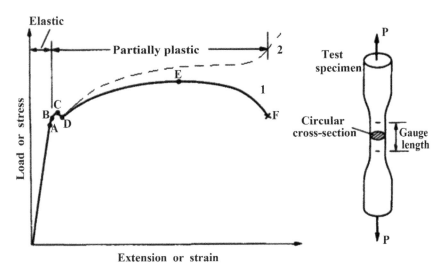

FIGURE 1.1   Typical tensile test curve for mild steel.

Some point A is eventually reached, however, when the linear nature of the graph ceases and this point is termed as a "limit of proportionality." For a short period beyond this point, the material may still be elastic in the sense that deformations are completely recovered when load is removed (i.e., strain returns to zero) but Hooke's law does not apply. The limiting point B for this condition is termed the elastic limit. For most practical purposes, it can often be assumed that points A and B are coincident.

Beyond the elastic limit, plastic deformation occurs and strains are not totally recoverable. There will thus be some permanent deformation or permanent set when load is removed. After the point C, termed the upper yield point, and D, the lower yield point, relatively rapid increases in strain occur without correspondingly high increases in load or stress. The graph thus becomes much more shallow and covers a much greater portion of the strain axis than does the elastic range of the material. The capacity of a material to allow these large plastic deformations is a measure of the so-called ductility of the material. Beyond the yield point some increase in load is required to take the strain to point E on the graph. Between D and E, the material is said to be in the elastic-plastic state, some of the section remaining elastic and hence contributing to recovery of the original dimensions if load is removed, the remainder being plastic. Beyond E the cross-sectional area of the bar begins to reduce rapidly over a relatively small length of the bar and the bar is said to neck. This necking takes place whilst the load reduces, and fracture of the bar finally occurs at point F. The nominal stress at failure, termed the maximum or ultimate tensile stress, is given by the load at E divided by the original cross-sectional area of the bar. (This is also known as the tensile strength of the material of the bar.) Owing to the large reduction in area produced by the necking process, the actual stress at fracture is often greater than the above value. Since, however, designers are interested in maximum loads which can be carried by the complete cross-section, the stress at fracture is seldom of any practical value.

Many engineering components are exposed to high temperature for a long period of time changes within the component due to this (at constant stress) is called creep, for example, turbine blade within a jet engine, steam generator, steel power plants, oil refineries, rubber band, concrete bridge deck: sagging between supports due to self-weight therefore the deck is constructed with an upward camber, and so on.

Creep is a progressive time-dependent plastic deformation, which generally occurs at high temperatures $(T > (0.3 - 0.6)T_m)$, under a constant load or stress[8] where T_m is a material's absolute melting temperature and T sometimes called thermal creep for metals $T > 0.4T_m$ and for ceramics $T > 0.5T_m$. It can also happen at room temperature for soft metals such as lead or glass even it can also happen at much slower temperature such as albeit.

The tests which are used to measure elevated temperature strength must be selected on the basis of the time scale of the service which the material must withstand. The creep test measures the dimensional changes which occur from elevated temperature exposure.

To determine the engineering creep curve of a metal, a constant load is applied to a tensile specimen maintained at a constant temperature, and the strain (extension) of the specimen is determined as a function of time. Although the measurement of creep resistance is quite simple in principle, in practice it requires considerable laboratory equipment.

Curve A, obtained from creep test in Figure 1.2, illustrates the idealized shape of a creep curve. The slope of this curve ($d\varepsilon/dt = \dot{\varepsilon}$) is referred to as the creep rate. Following an initial rapid elongation of the specimen, ε_0, the creep rate decreases with time, then reaches essentially a steady state in which the creep rate changes little with time, and finally the creep rate increases rapidly with time until fracture occurs.

Thus, it is natural to discuss the creep curve in terms of its three stages. It should be noted, however, that the degree to which these three stages are readily distinguishable depends strongly on the applied stress and temperature. In making an engineering creep test, it is usual practice to maintain the load constant throughout the test. Thus, as the specimen elongates and decreases in cross-sectional area, the axial stress increases. The initial stress, which was applied to the specimen, is usually the reported value of stress. Methods of compensating for the change in dimensions of the specimen so as to carry out the creep test under constant stress conditions have been developed. When constant stress tests are made, it is frequently found that no region of accelerated creep rate occurs (region III, Fig. 1.2) and a creep curve similar to B in Figure 1.2 is obtained. Accelerated creep is found, however, in constant stress tests when metallurgical changes occur in the metal. Curve B should be considered representative of the basic creep curve for a metal. According to the idealized Curve A, the three stages of creep are shown in Figure 1.2:

a) Stage 1: primary creep (transient creep stage): Primary creep is a period of predominantly transient creep in which the creep resistance of the material increases by virtue of its own deformation. For low temperatures and stresses, as in the creep of lead at room temperature, primary creep is the predominant creep process. Strain rate decreases as strain increases and slows with time and it has resistance to plastic deformation: strain hardening as shown in Figure 1.3.

b)   Stage 2: secondary (steady state) creep: Secondary creep is a period of nearly constant creep rate which results from a balance between the competing processes of strain hardening and recovery. It is used as a design tool; strain rate is minimum and constant and fracture will not occur as shown in Figure 1.3.

c)   Stage 3: tertiary creep (accelerating creep stage): Tertiary creep is a period mainly occurs in constant load creep tests at high stresses at high temperatures. The reasons for the accelerated creep rate which leads to rapid failure are not well known. Strain rate increases (accelerating) and terminates when the materials rupture, reduction in cross-sectional due to voids, necking reduce, failure as shown in Figure 1.3.

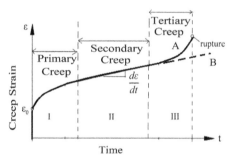

FIGURE 1.2   Typical creep curve, Curve A, constant load test, Curve B, constant stress test.

Andrade's (1957) showed that the constant stress–creep curve represents the superposition of two separate creep processes, which occur after the sudden straining due to application of load. The first component of the creep curve is a transient creep in which the creep rate decreases with time. Added to this is a constant rate viscous creep component. The superposition of these creep processes is shown in Figure 1.4. Andrade observed that a creep curve can be represented by the following empirical equation:

$$\varepsilon = \varepsilon_0 \left(1 + \beta_1 t^{1/3}\right) e^{kt} \tag{1.1}$$

where ε_0 is the instantaneous strain, ε is the strain in time t, and β_1 and k are constants.

To analyze creep in engineering structures and components, mathematical models of creep are needed. For this purpose, the data obtained from the tensile tests serve as the main source of information about creep.

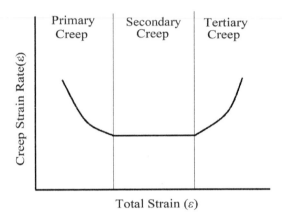

FIGURE 1.3   Creep strain rate.

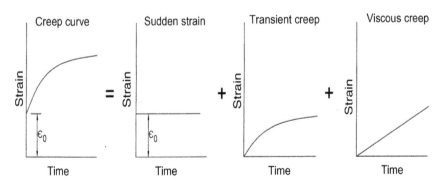

FIGURE 1.4   Analysis of the competing processes determining the creep curve (Andrade, 1957).

1.1.1.4   *ELASTIC-PLASTIC AND CREEP TRANSITIONS PHENOMENON*

Elastic-plastic and creep transitions are well-known examples of natural phenomenon that changes from one state into another. In the classical theory of plasticity, the material region is assumed to be divided into an elastic and plastic region, which is separated by a yield surface depending on the symmetry and other physical considerations such as yield condition for elastic-plastic deformation, and yield condition and creep laws for creep deformation. The main drawbacks of the classical treatment are the following assumptions:

a) Even though the material at a point has yielded, the material at a neighboring point still remains elastic.
b) A yield surface of the assumed type separating the elastic and plastic regions exist.
c) For any given material, there exists a function of the three principal stresses which always has a value when yielding begins regardless of the stress state.
d) The same functional relationship applies to all materials although the numerical value of the function is different for different materials.

In the classical theory of elasticity, in particular, the displacements are assumed to be so small that the squares and products of displacement gradients are neglected and the measure of strain thus becomes linear. But nonlinear terms are very important at transition state. Linearization of problems has all the advantages of existence, uniqueness, and stability. But it also has disadvantages in that it may not be able to explain or represent all the changes and phenomena occurring in a medium.

The classical theory of elasticity and plasticity, as has already been mentioned, divides the spectrum of deformation of solids into two distinct states, one in which the deformation is recoverable and the other in which it is not. Both elastic and plastic field equations are solved separately and later joined together by the so-called yield condition. Perfect elasticity and ideal plasticity are two extreme properties of the material and the use of an ad-hoc rule like yield condition amounts to divide the two extreme properties by a sharp line which is not physically possible. When a material passes from one state to another qualitatively different state, transition takes place. Since this transition is nonlinear in character and difficult to investigate, workers have taken certain ad-hoc assumptions like yield condition, incompressibility condition and a strain law, which may or may not valid for the problem. It is therefore natural to expect that any physically realistic theory should include mid-zone or transition state.

A medium cannot change from state A into state B without passing through an intermediate state T. In a large number of cases A and B may be treated as linear fields, but T is essentially a nonlinear field, since both A and B dovetail into each other in T. Transition fields are, therefore, nonlinear, nonconservative, and irreversible in nature and should not be treated as superposition of effects. This essentially nonlinear transition region T remains untreated by research workers and instead nondifferentiable, singular surfaces are introduced to connect the regions A and B. These, in turn have necessitated the increasing use of ad-hoc, semi-empirical laws, such as yield conditions, creep-strain laws, jump conditions across shocks, and so on. This ad-hoc, semi-empirical laws are based on long experimental results, but very few authors have ever tried to justify these ad-hoc assumptions with analytical basis. Seth's approach toward this end is an attempt at bridging the gap between seemingly unrelated phenomena, the approach which is based on sound mathematical ground. Seth[89] has given a thorough analytical treatment of the intermediate region. He has named this intermediate region as "Transition Region." So A passes into B through T.

In a series of papers, Seth[86–91] has given an entirely different orientation to this interesting problem of transition. Seth has argued that at transition, the differential system governing the physical phenomenon should attain some sort of criticality. Once the "critical points" or "Transition points" are recognized, the asymptotic solutions at these "Transition" points give the solutions corresponding to the "Transition" states. He has developed a new "Transition theory" of elastic-plastic and creep deformation on sound analytical base. This theory is applied to many research problems related to disc (solid and annular), cylinder (thin and thick), shells and plates, and so on.

1.1.1.4.1 Identification of the Transition State

When a material at a point has yielded, it is more reasonable to expect that the material at the neighboring points are on their way to yield, rather than assume that they remain in the elastic state as completely opposed to the plastic state of the nearby particle. As the plastic yielding of a material is a consequence of collapse of its internal or macroscopic structure, the plastic yielding will be complete or partial depending on the existing physical conditions. This leads us again to the recognition of two material states: a transition state and a plastic state. There are three ways in which a transition could be identified analytically: three different ways to explain how transition may occur from one state to another state analytically:

1. At transition, the differential system defining the elastic state should attain some criticality.
2. The complete breakdown of the macroscopic structure at transition should correspond to the degeneracy of the material (spatial) strain ellipsoid. This means that the length of at least one of the axes of the strain ellipsoid should be zero or infinity.
3. If we consider the plastic state B as an image of the elastic state A, under the transformation $x^k = x^k (X^k)$, then at transition, the Jacobian value of the transformation would be zero or infinity. This means that when transition occurs, one to one correspondence between A and B no longer holds.

Borah[5] identified that all the three treatments lead to the same result and (3) is the most general form of yield condition. At transition, the nature of this group changes. For example, an elastic body, which belongs to the orthogonal group, becomes unimolecular on becoming plastic. Lastly, from the macro point of view, one can imagine that at transition the macro-element breaks down, with the result that the corresponding transformation matrix becomes singular.

In general, any deformable medium under an external loading system has three states: (1) elastic, (2) transition, and (3) plastic (or creep) that means the material from elastic state can go over into plastic state, or to creep state, or first to plastic then to creep or vice-versa, under external loading system. When the material under experiment goes from primary state of creep to secondary or to tertiary states and secondary to tertiary state, transition takes place. A plastic or a creep state is a transition state from initially elastic one. All these cases of transition can be expected to occur when some functions of elasticity of the medium take on critical values. These functions are called transition functions. These may be either principal stresses or principal stress difference, or stress invariant or may be any suitable combination of these. These critical (asymptotic) values have to be determined at the transition points of the differential equations describing the medium. If a number of transition states occur at the same point, the transition function will have different limiting values, and the point will be a multiple one, each branch of which will then correspond to different states, where as the classical treatment uses different constitutive equations for each state involved. But in nonclassical treatment, there does exist a constitutive equation governing the transition state and it may be obtained from the elastic state with an asymptotic approach. The constitutive equation for the plastic state will follow in a similar manner from the transition state.

1.1.2   *FINITE AND INFINITESIMAL DEFORMATIONS*

Deformation in the classical theory of elasticity is named infinitesimal for arbitrary particle of the medium when the spatial derivatives of the components of the displacement vector are so small that their products and squares may be neglected. If the displacement gradient components are small, then finite strain components reduce to infinitesimal strain components and these components represent small deformations. Therefore, the infinitesimal theory is also known as small deformation theory. Many applications of theory of plasticity involve large deformation for which infinitesimal theory is not adequate. Many trials have been attempted to extend the classical theory of infinitesimal deformations to the case of finite deformations but elementary hypothesis which defines that infinitesimal deformations are not reasonable to finite deformation. Moreover, for large rotation problems, infinitesimal theory is insufficient, even if straining and stretching is small. The finite deformation theory[110] has been applied to solve different problems of material mechanics for which classical theory of infinitesimal deformation is not adequate. In finite deformation, displacements and its derivatives components are not linear, whereas in infinitesimal deformation displacement and its derivative, components are linear. Two theories of elastic failure have been combined, that is, principal maximum stress theory and maximum shear stress to predict a yield point. In addition to this, it has also demonstrated that yield stress in tension is less as compared with yield stress in compression, which is in agreement with Bauschinger effect. The applications of finite deformation theory give longitudinal stresses in cylinders subjected to twist, which are ignored in the infinitesimal theory. To explore finite and infinitesimal deformations, two principal approaches have been used in this present chapter, that is, classical theory and transition theory.[87] Infinitesimal deformations have been considered in classical theory while finite deformations have been considered in transition theory.

To explore finite deformations in continuous media, two methods are usually used, namely, Lagrangian and Eulerian methods. The components of strain in these two methods are described by the coordinates of a particle either in the initial (unstrained) state or in the final (strained) state as independent variables. Due to mathematical convenience of the Lagrangian method many researchers adopted this method, whereas Seth[85] and many other authors focused on the use of the Eulerian method. For infinitesimal deformations, the Eulerian and Lagrangian presented the same aspect. Therefore, no issues were there regarding distinction between these two methods for small deformations.

Consider an aggregate of particles in a continuous medium. Let $X = (X_1, X_2, X_3)$ be the coordinates of a particle lying on a curve C_0 containing the vector PQ before deformation and let $x = (x_1, x_2, x_3)$ be the coordinates of the same particle lying on a curve C_1 containing the vector $P'Q'$ after deformation which are shown in Figure 1.5. Let dS be the length of the vector PQ (i.e., dX) and ds be the length of $P'Q'$ (i.e., dx). Then

$$\left(dS\right)^2 = dX \cdot dX = dX_1^2 + dX_2^2 + dX_3^2 = dX_i dX_i, \quad \left(i = 1,2,3\right) \qquad (1.2)$$

and

$$\left(ds\right)^2 = dx \cdot dx = dx_1^2 + dx_2^2 + dx_3^2 = dx_i dx_i, \quad \left(i = 1,2,3\right) \qquad (1.3)$$

We can write

$$dX_i = X_{i,j} dx_j$$

In *Eulerian* description (spatial description) of strain, we have

$$X_i = X_i\left(x_1, x_2, x_3\right) \qquad (1.4)$$

Substituting equation (1.4) in equation (1.2)

$$(dS)^2 = dX_i dX_i = X_{i,j} X_{i,k} dx_j dx_k \quad (i,j,k = 1,2,3) \qquad (1.5)$$

and

$$(ds)^2 = dx_i dx_i = \delta_{jk} dx_j dx_k \quad (i,j,k = 1,2,3) \qquad (1.6)$$

The necessary and sufficient condition for the transformation $X_i = X_i$ (x_1, x_2, x_3) to be one of rigid body motion is that $(dS)^2$ and $(ds)^2$ should be equal for all curves C_0. Hence, we take the difference $(ds)^2 - (dS)^2$ as a measure of strain and write

$$(ds)^2 - (dS)^2 = 2e_{jk}^A dx_j dx_k \qquad (1.7)$$

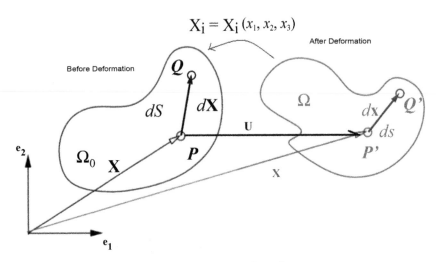

FIGURE 1.5 Body, its reference, and present configurations.

The strain tensor e_{jk}^A was introduced by Cauchy for infinitesimal strains and by Almansi and Hamel finite strain and is known as Almansi strain tensor.

Subtracting equation (1.5) from equation (1.6), we have,

$$(ds)^2 - (dS)^2 = \left(\delta_{jk} - X_{ij}X_{ik}\right)dx_j dx_k \tag{1.8}$$

Comparing equation (1.7) with equation (1.8), we get

$$2e_{jk}^A = \delta_{jk} - X_{ij}X_{ik} \tag{1.9}$$

We can write strains e_{jk}^A in terms of displacement components $u_i = x_i - X_i$ as

$$2e_{jk}^A = \left[u_{j,k} + u_{k,j} - u_{i,j}u_{i,k}\right] \tag{1.10}$$

where the function e_{jk}^A are called the *Eulerian* strain components.

If the Lagrangian coordinates e_{jk}^A are taken as independent variable and equations of transformation are of the form $x_i = x_i\,(X_1,X_2,X_3)$
We can write

$$dx_i = x_{i,j}dX_j \quad (i,j=1,2,3) \tag{1.11}$$

Thus the elements dS and ds of the arc of the curve C_0 and C_1 are given by

$$(dS)^2 = dx_i dx_i = \delta_{jk} dx_j dx_k \qquad (1.12)$$

and

$$(ds)^2 = dx_i dx_i = x_{i,j} x_{i,k} dx_j dx_k \qquad (1.13)$$

The Lagrangian components of strain ε_{jk} are defined as

$$(ds)^2 - (dS)^2 = 2\varepsilon_{jk} dX_j dX_k \qquad (1.14)$$

Also from equation (1.12) and (1.13), we have

$$(ds)^2 - (dS)^2 = \left(x_{i,j} x_{i,k} - \delta_{jk} \right) dX_j dX_k \qquad (1.15)$$

Now comparing equation (1.14) and (1.15), we get

$$2\varepsilon_{jk} = x_{i,j} x_{i,k} - \delta_{jk} \qquad (1.16)$$

We can express ε_{jk} in terms of displacement components $u_i = x_i - X_i$ as,

$$2\varepsilon_{jk} = \left[u_{j,k} + u_{k,j} + u_{i,j} u_{i,k} \right]; j,k = 1,2,3 \qquad (1.17)$$

Let $u_1 = u$, $v_1 = v$, $u_3 = w$, $x_1 = x$, $x_2 = y$, $x_3 = z$, $X_1 = X$, $X_2 = Y$, and $X_3 = Z$.
To show the fact that the differentiation in equation (1.10) is carried out with respect to the variable x_i, while equation (1.17) is carried out with respect to the variable X_i, the typical expressions for e_{jk}^A and ε_{jk} in an *unbridged* form can be written as

$$e_{xx}^A = \frac{\partial u}{\partial x} - \frac{1}{2}\left[\left(\frac{\partial u}{\partial x}\right)^2 + \left(\frac{\partial v}{\partial x}\right)^2 + \left(\frac{\partial w}{\partial x}\right)^2 \right]$$

$$\varepsilon_{XX} = \frac{\partial u}{\partial X} + \frac{1}{2}\left[\left(\frac{\partial u}{\partial X}\right)^2 + \left(\frac{\partial v}{\partial X}\right)^2 + \left(\frac{\partial w}{\partial X}\right)^2 \right]$$

$$2e_{xy}^A = \frac{\partial u}{\partial y} + \frac{\partial v}{\partial x} - \left[\frac{\partial u}{\partial x}\frac{\partial u}{\partial y} + \frac{\partial v}{\partial x}\frac{\partial v}{\partial y} + \frac{\partial w}{\partial x}\frac{\partial w}{\partial y} \right]$$

$$2\varepsilon_{XY} = \frac{\partial u}{\partial Y} + \frac{\partial v}{\partial X} + \left[\frac{\partial u}{\partial X}\frac{\partial u}{\partial Y} + \frac{\partial v}{\partial X}\frac{\partial v}{\partial Y} + \frac{\partial w}{\partial X}\frac{\partial w}{\partial Y} \right]$$

where e_{xx}^A and ε_{XX} represent the extension of vectors originally parallel to the coordinate axis, while e_{xy}^A and ε_{XY} represent shear or change of angle between vectors originally at the right angles. The classical theory supports infinitesimal deformation which is also known as small deformation theory. The basic requirement in classical theory is that displacement gradient should be less than unity if $\dfrac{\partial u}{\partial x} \ll 1$, $\dfrac{\partial u}{\partial x} \ll 1$, then difference between Lagrangian and *Eulerian* will disappear and both will coincide and linear. Therefore, equations (1.10) and (1.17) can be expressed as

$$\varepsilon_{ij} = e_{ij}^A = \frac{1}{2}\left[u_{i,j} + u_{j,i}\right] \quad (i,j = 1,2,3) \tag{1.18}$$

The strain components in equation (1.10) in cylindrical polar coordinates are given by

$$e_{rr}^A = \frac{\partial u}{\partial r} - \frac{1}{2}\left[\left(\frac{\partial u}{\partial r}\right)^2 + \left(\frac{\partial v}{\partial r}\right)^2 + \left(\frac{\partial w}{\partial r}\right)^2 - v^2\right]$$

$$e_{\theta\theta}^A = \frac{1}{r}\frac{\partial v}{\partial \theta} + \frac{u}{r} - \frac{1}{2r^2}\left[\left(\frac{\partial u}{\partial \theta}\right)^2 + r^2\left(\frac{\partial v}{\partial \theta}\right)^2 + \left(\frac{\partial w}{\partial \theta}\right)^2\right]$$

$$-\frac{1}{2r^2}\left[-vr^2\frac{\partial u}{\partial \theta} + u\frac{\partial v}{\partial \theta} - v\frac{\partial u}{\partial \theta} + ur^2\frac{\partial v}{\partial \theta} + u^2 + r^2v^2\right]$$

$$e_{zz}^A = \frac{\partial w}{\partial z} - \frac{1}{2}\left[\left(\frac{\partial u}{\partial z}\right)^2 + r\left(\frac{\partial v}{\partial z}\right)^2 + \left(\frac{\partial w}{\partial z}\right)^2\right]$$

$$e_{r\theta}^A = \frac{1}{2}\left[\frac{1}{r}\frac{\partial u}{\partial \theta} + \frac{\partial v}{\partial r} - \frac{u}{r}\right] - \frac{1}{2r}\left[\frac{\partial u}{\partial r}\frac{\partial u}{\partial \theta} + r\frac{\partial(rv)}{\partial r}\frac{\partial v}{\partial \theta} + \frac{\partial w}{\partial r}\frac{\partial w}{\partial \theta}\right]$$

$$-\frac{1}{2r}\left[-vr^2\frac{\partial u}{\partial r} + \frac{u}{r}\frac{\partial(rv)}{\partial r} - rv\frac{\partial v}{\partial \theta}\right]$$

$$e_{\theta z}^A = \frac{1}{2}\left[\frac{\partial v}{\partial z} + \frac{1}{r}\frac{\partial w}{\partial \theta}\right] - \frac{1}{2r}\left[\frac{\partial u}{\partial \theta}\frac{\partial u}{\partial z} + r^2\frac{\partial v}{\partial \theta}\frac{\partial v}{\partial z} + \frac{\partial w}{\partial \theta}\frac{\partial w}{\partial z}\right]$$

$$-\frac{1}{2r}\left[-v\frac{\partial u}{\partial z} + r^2 u\frac{\partial v}{\partial z}\right]$$

$$e_{zr}^A = \frac{1}{2}\left[\frac{\partial w}{\partial r} + \frac{\partial u}{\partial z}\right] - \frac{1}{2r}\left[\frac{\partial u}{\partial z}\frac{\partial u}{\partial r} + r\frac{\partial v}{\partial z}\frac{\partial(rv)}{\partial r} + \frac{\partial w}{\partial z}\frac{\partial w}{\partial r} + rv\frac{\partial v}{\partial z}\right]$$

where u, v, w are the physical components of displacement u_i and e_{rr}^A, $e_{\theta\theta}^A$ e_{zz}^A, $e_{r\theta}^A$ $e_{\theta z}^A$, e_{zr}^A are the components of the strain tensor e_{ij}^A.

For infinitesimal deformation, the strain components are

$$e_{rr}^A = \frac{\partial u}{\partial r}, \quad e_{\theta\theta}^A = \frac{1}{r}\frac{\partial v}{\partial \theta} + \frac{u}{r}, \quad e_{zz}^A = \frac{\partial w}{\partial z}$$

For instance, on account of symmetry of the problem, we may assume the following displacements:

$u = r(1-\beta)$ $v = 0$, $w = dz$ where β is a function of r and $r = \sqrt{x^2 + y^2}$. Then, the *Almansi* strain measures are

$$e_{rr}^A = \frac{\partial u}{\partial r} = \frac{\partial(r(1-\beta))}{\partial r} = 1 - (\beta + r\beta')$$

$$e_{\theta\theta}^A = \frac{1}{r}\frac{\partial v}{\partial \theta} + \frac{u}{r} = 1 - \beta$$

$$e_{zz}^A = \frac{\partial w}{\partial z} = d$$

$$e_{r\theta}^A = e_{\theta z}^A = e_{zr}^A = 0$$

where $\beta' = d\beta/dr$.

1.1.3 GENERALIZED STRAIN MEASURE

The response of all materials to the external loading, in general, is nonlinear in character. The division of deformation into different states is the result of linearization of engineering problems. In case of large deformations such as plastic flow, creep, and fatigue, the classical treatment requires a number of ad-hoc and semi-empirical laws. The use of these semi-empirical laws has made the problem complicated without evolving any simple concept governing them. One source of complications is the use of classical measures of deformation produced in a medium even when we know that nonlinearity is a characteristic of such deformed media. In classical mechanics, ordinary measures are found sufficient and therefore there is no need to extend them. The equation of equilibrium and the concept of stresses are well defined, but the measures of deformation are flexible. A continuous approach requires the introduction of nonlinear measures. Deformation fields associated with irreversible phenomenon such as elastic-plastic deformations, creep, relaxation, fatigue and fracture, and so on are nonlinear in character as explained by extensive experimental studies. The classical measures of deformation are not sufficient to deal with transitions and hence the corresponding constitutive equations of the medium are complicated. On the basis of above analysis, it is found that we should have to construct generalized measure of deformation to overcome the difficulty of using classical measures in continuum mechanics.

Rabotnov[78,79] has pointed out the ambiguities of the experimental data which involves in the choice of suitable constitutive equations for different states of creep described in Figure 1.2. There are two parameters which characterize each of these states; one is for the measure and other for the irreversibility. Therefore, it is expected that a generalized measure concept in which these two parameters are experimentally determined may give better results in creep deformation. The strain-rate described in four stages of creep elastic, transient, secondary, and rupture is different in each case (Fig. 1.2). As the deformations are nonlinear, there arises the need of generalization of strain rate measure so that they can be used in all the stages of creep.

Seth[92] has defined the generalized principal strain measure as

$$e_{ii} = \int_{0}^{e_{ii}^{A}} \left(1 - 2e_{ii}^{A}\right)^{(n/2)-1} \mathrm{d}e_{ii}^{A} = \frac{1}{n}\left(1 - \left(1 - 2e_{ii}^{A}\right)^{n/2}\right) \tag{1.19}$$

where n is the coefficient of strain measure, e_{ii}^{A} is the principal Almansi finite strain component, and $i = 1,2,3$.

In Cartesian coordinates, we can write down the generalized measure in terms of any other measures.

For uniaxial case, it is given by Seth[93] as

$$e = \frac{1}{n}\left[1 - \left(\frac{l_0}{l}\right)^n\right] \tag{1.20}$$

where l_0 and l are the initial and strained lengths of the rod respectively. For $n = 0,1,2, -1, -2$, it gives the Hencky, Swainger, Almansi, Cauchy, and Green measures, respectively. Seth[92,94,95] has shown that the well-known creep strain laws used in current literature such as Norton's law, Kachonov's law, Odqvist's law, Andrade's law, and so on can be derived from the generalized measure.

The generalized strain measure not only gives many well-known strain measures for different values of n, but it is also used to find the creep stresses when combined with the transition points of the governing differential equations. Seth has also shown that the transition point analysis does not require any assumptions of incompressibility, creep strain law, and yield conditions like in classical theory. He successfully applied transition theory to a large number of problems. The asymptotic solution of the governing differential equations at the transition point gives the same results which are also obtained by assuming yield criteria when they exist. The most important contribution of these generalized measures is that they eliminate the use of semi-empirical laws and jump conditions. If such laws exist, they come out with the analytic treatment as a particular case. Also the important feature of nonlinear measure is that, they explain transition without assuming any conditions to match the two solutions at transition.

1.1.4 CONSTITUTIVE EQUATIONS

The necessary set of equations required in engineering mechanics is strain-displacement equation, constitutive equation, equilibrium equation, and compatibility equation. Figure 1.6 pictorially depicts the concepts that these equations relate. Thus, the strain displacement relation allows one to compute the strain given a displacement; constitutive relation gives the value of stress for a known value of the strain or vice versa; equilibrium equation, crudely, relates the stresses developed in the body to the forces and moment applied on it; and finally compatibility equation places restrictions on how the strains can vary over the body so that a continuous displacement field could be found for the assumed strain field.

Equations characterizing the individual material and its response to applied loads are called constitutive equations. The macroscopic behavior is described by these equations resulting from the internal constitution of the material. Materials in the solid state behave in such a complex way that when entire range of possible temperatures and deformations is considered, it is not feasible to write down one equation or set of equations to describe accurately a real material over its entire range of behavior. Instead, separate equations are formulated to describe the various kinds of ideal materials response, each of which is a mathematical formulation designed to approximate physical observations of a real material's response over a suitably restricted range. The classical equations were introduced separately to meet specific needs and made as simple as possible to describe many physical situations. Some of the ideas involved in formulating simple equations for such ideal material are illustrated below.

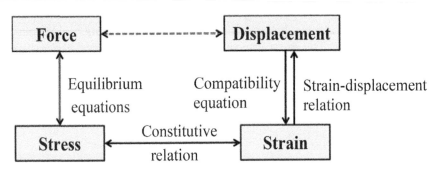

FIGURE 1.6   Basic concepts and equations in mechanics.

1.1.4.1   *ELASTIC STATE*

A material is said to be in perfect elastic state when it recovers its original shape completely after removal of load causing the deformation. Also, there is one to one relationship between the state of stress and state of strain at a given temperature and material obeys Hooke's law, which states that stresses are proportional to strains. Under triaxial loading, classical theory of elasticity assumes that generalized Hooke's law expresses each stress component as a linear combination of strains, that is

$$\sigma_{ij} = C_{ijkl}\varepsilon_{kl} \quad (i,j,k = 1,2,3) \tag{1.21}$$

where σ_{ij}, ε_{kl}, and C_{ijkl} are the stress tensor, strain tensor, and stiffness components, respectively. These nine equations contain 81 material constants C_{ijkl}, but not all the constants are independent. Due to the symmetry of σ_{ij} and ε_{kl}, these 81 material constants reduce to only 36 material constants and due to strain energy density, it reduces to 21 material constants. Therefore, a material in which the elastic properties depend on the orientation of the sample is called anisotropic. If the anisotropic elasticity of the material is to be fully described, a total of 21 independent parameters are needed in six by six stiffness matrix C_{ijkl}, as follows:

$$\begin{bmatrix} \sigma_1 \\ \sigma_2 \\ \sigma_3 \\ \sigma_4 \\ \sigma_5 \\ \sigma_6 \end{bmatrix} = \begin{bmatrix} C_{11} & C_{12} & C_{13} & C_{14} & C_{15} & C_{16} \\ C_{12} & C_{22} & C_{23} & C_{24} & C_{25} & C_{26} \\ C_{13} & C_{23} & C_{33} & C_{34} & C_{35} & C_{36} \\ C_{14} & C_{24} & C_{34} & C_{44} & C_{45} & C_{46} \\ C_{15} & C_{25} & C_{35} & C_{45} & C_{55} & C_{56} \\ C_{16} & C_{26} & C_{36} & C_{46} & C_{56} & C_{66} \end{bmatrix} \begin{bmatrix} \varepsilon_1 \\ \varepsilon_2 \\ \varepsilon_3 \\ \varepsilon_4 \\ \varepsilon_5 \\ \varepsilon_6 \end{bmatrix} \qquad (1.22)$$

In the case of orthotropic materials (materials which possess three orthogonal planes of elastic symmetry) the number of elastic constants is further reduced to nine.

$$\begin{bmatrix} \sigma_1 \\ \sigma_2 \\ \sigma_3 \\ \sigma_4 \\ \sigma_5 \\ \sigma_6 \end{bmatrix} = \begin{bmatrix} C_{11} & C_{12} & C_{13} & 0 & 0 & 0 \\ C_{12} & C_{22} & C_{23} & 0 & 0 & 0 \\ C_{13} & C_{23} & C_{33} & 0 & 0 & 0 \\ 0 & 0 & 0 & C_{44} & 0 & 0 \\ 0 & 0 & 0 & 0 & C_{55} & 0 \\ 0 & 0 & 0 & 0 & 0 & C_{66} \end{bmatrix} \begin{bmatrix} \varepsilon_1 \\ \varepsilon_2 \\ \varepsilon_3 \\ \varepsilon_4 \\ \varepsilon_5 \\ \varepsilon_6 \end{bmatrix} \qquad (1.23)$$

Some examples of orthotropic materials are wood, two-ply fiber-reinforced composites and human bones, and so on.

An orthotropic material is called transversely isotropic when one of its principal planes is a plane of isotropy, that is, at every point there is a plane on which the mechanical properties are the same in all directions. In this case, there are only five independent elastic constants.

$$\begin{bmatrix} \sigma_1 \\ \sigma_2 \\ \sigma_3 \\ \sigma_4 \\ \sigma_5 \\ \sigma_6 \end{bmatrix} = \begin{bmatrix} C_{11} & C_{12} & C_{12} & 0 & 0 & 0 \\ C_{12} & C_{22} & C_{23} & 0 & 0 & 0 \\ C_{12} & C_{23} & C_{22} & 0 & 0 & 0 \\ 0 & 0 & 0 & (C_{22}-C_{23})/2 & 0 & 0 \\ 0 & 0 & 0 & 0 & C_{55} & 0 \\ 0 & 0 & 0 & 0 & 0 & C_{55} \end{bmatrix} \begin{bmatrix} \varepsilon_1 \\ \varepsilon_2 \\ \varepsilon_3 \\ \varepsilon_4 \\ \varepsilon_5 \\ \varepsilon_6 \end{bmatrix} \qquad (1.24)$$

Some examples of transversely isotropic material are fiber-reinforced composites, magnesium, beryl, cadmium, and so on.

An isotropic material is characterized by an infinite number of planes of material symmetry through a point. Thus, an isotropic material is fully characterized by only two independent constants, for example, the *stiffnesses* C_{11} and C_{12}:

$$\begin{bmatrix} \sigma_1 \\ \sigma_2 \\ \sigma_3 \\ \sigma_4 \\ \sigma_5 \\ \sigma_6 \end{bmatrix} = \begin{bmatrix} C_{11} & C_{12} & C_{12} & 0 & 0 & 0 \\ C_{12} & C_{11} & C_{12} & 0 & 0 & 0 \\ C_{12} & C_{12} & C_{11} & 0 & 0 & 0 \\ 0 & 0 & 0 & (C_{11}-C_{12})/2 & 0 & 0 \\ 0 & 0 & 0 & 0 & (C_{11}-C_{12})/2 & 0 \\ 0 & 0 & 0 & 0 & 0 & (C_{11}-C_{12})/2 \end{bmatrix} \begin{bmatrix} \varepsilon_1 \\ \varepsilon_2 \\ \varepsilon_3 \\ \varepsilon_4 \\ \varepsilon_5 \\ \varepsilon_6 \end{bmatrix} \quad (1.25)$$

In terms of Lame's constants λ and μ, these constants can be written as $C_{12} = \lambda$ and $(C_{11} - C_{12})/2 = \mu$. Then, $C_{11} - \lambda + 2\mu$.

By performing simple tensile and shear tests on orthotropic materials, all of the components C_{ij} of the stiffness matrix can be related to physical or mechanical properties of the material. Thus,

$$\begin{bmatrix} \sigma_{11} \\ \sigma_{22} \\ \sigma_{33} \\ \sigma_{23} \\ \sigma_{31} \\ \sigma_{12} \end{bmatrix} = \begin{bmatrix} \dfrac{1-v_{23}v_{32}}{E_2 E_3 \Delta} & \dfrac{v_{21}+v_{31}v_{23}}{E_2 E_3 \Delta} & \dfrac{v_{31}+v_{21}v_{32}}{E_2 E_3 \Delta} & 0 & 0 & 0 \\ \dfrac{v_{12}+v_{13}v_{32}}{E_3 E_1 \Delta} & \dfrac{1-v_{31}v_{13}}{E_3 E_1 \Delta} & \dfrac{v_{32}+v_{31}v_{12}}{E_3 E_1 \Delta} & 0 & 0 & 0 \\ \dfrac{v_{13}+v_{12}v_{23}}{E_1 E_2 \Delta} & \dfrac{v_{23}+v_{13}v_{21}}{E_1 E_2 \Delta} & \dfrac{1-v_{12}v_{21}}{E_1 E_2 \Delta} & 0 & 0 & 0 \\ 0 & 0 & 0 & G_{23} & 0 & 0 \\ 0 & 0 & 0 & 0 & G_{31} & 0 \\ 0 & 0 & 0 & 0 & 0 & G_{12} \end{bmatrix} \begin{bmatrix} \varepsilon_{11} \\ \varepsilon_{22} \\ \varepsilon_{33} \\ \varepsilon_{23} \\ \varepsilon_{31} \\ \varepsilon_{12} \end{bmatrix} \quad (1.26)$$

where $\Delta = \left(\left(1 - v_{12}v_{21} - v_{23}v_{32} - v_{31}v_{13} - 2v_{12}v_{23}v_{31}\right)/E_1 E_2 E_3 \right)$, γ_{ij} is an engineering shear strain, ε_{ij} is the tensorial shear strain, where $i \neq j$ such that $\varepsilon_4 = \gamma_{23} = 2\varepsilon_{23}$, $\varepsilon_5 = \gamma_{31} = 2\varepsilon_{31}$, $\varepsilon_5 = \gamma_{12} = 2\varepsilon_{12}$, and E_1, E_2, and E_3 are the Young's moduli in the principal directions. The Poisson's ratio, v_{ij} is defined as $v_{ij} = -\varepsilon_{jj}/\varepsilon_{ii}$ (no summations).

For an isotropic elastic material in which there is no change of temperature, Hooke's law referred to a set of rectangular Cartesian coordinates may be stated in the form:

$$\sigma_{\alpha\alpha} = 3K e_{\alpha\alpha}, \quad (1.27)$$

$$\sigma'_{ij} = 2G e'_{ij} \quad (1.28)$$

where K and G are the elastic constants bulk modulus and shear modulus, respectively, and the primes denote the stress and strain deviators and

$$\sigma'_{ij} = \sigma_{ij} - \frac{\sigma_{\alpha\alpha}\delta_{ij}}{3}, \quad (1.29)$$

$$e'_{ij} = e_{ij} - \frac{e_{\alpha\alpha}\delta_{ij}}{3} \tag{1.30}$$

The coefficient 2 included in equation (1.28) makes e'_{ij} a tensor. Before the tensor concept was introduced, it was customary to define the shear strain as $\gamma'_{ij} = 2e'_{ij}$ or $\gamma_{ij} = 2e_{ij}$ for $i \neq j$. We have seen before that $\sigma_{\alpha\alpha}/3$ is the mean stress at a point and that, if the strain were infinitesimal, $e_{\alpha\alpha}$ is the change of volume per unit volume. Both $\sigma_{\alpha\alpha}$ and $e_{\alpha\alpha}$ are invariants. Thus, equation (1.27) states that the change of volume of the material is proportional to the mean stress. In the special case of hydrostatic compression, $\sigma_{xx} = \sigma_{yy} = \sigma_{zz} = -p$, $\sigma_{xy} = \sigma_{yz} = \sigma_{zx} = 0$, we have $\sigma_{\alpha\alpha} = -3p$, and equation (1.27) may be written, in the case of infinitesimal strain, with v and δv denoting volume and change of volume, respectively,

$$\frac{\delta v}{v} = \frac{-p}{K}, \tag{1.31}$$

Thus, the coefficient K is called the bulk modulus of the material. The strain deviation e'_{ij} describes a deformation without volume change. Equation (1.28) states that the stress deviation is simply proportional to the strain deviation. The constant G is called the modulus of elasticity in shear, or shear modulus, or the modulus of rigidity. In the special case of $e_{xy}, \sigma_{xy} \neq 0$, but all other strain and stress components vanish, we have

$$\sigma_{xy} = 2Ge_{xy} \tag{1.32}$$

If we substitute equations (1.29) and (1.30) into equation (1.28) and make use of equation (1.27), the result is the constitutive equations given by Hooke's stress–strain relation of the form:

$$\sigma_{ij} = \lambda e_{\alpha\alpha}\delta_{ij} + 2\mu e_{ij} \tag{1.33}$$

or

$$e_{ij} = \frac{\left[(1+v)\sigma_{ij} - v\sigma_{\alpha\alpha}\delta_{ij}\right]}{E} \tag{1.34}$$

where e_{ij} is the Almansi strain tensor, G is the shear modulus, $G = \mu$, $\lambda = K - 2/3G$, and δ_{ij} is Kronecker's delta. Even though we neglect the variation of the elastic constants with temperature, we may be compelled to take account of the thermal expansion of the material, which often produces dimensional changes as large as those produced by the applied forces, or, if the dimensional change is prevented by support constraints or surrounding material, thermal stresses are induced in addition to the stresses related to the strains according to the elastic constitutive equations.

The thermo-elastic constitutive equations for

a) isotropic material are given by[19,76]

$$\sigma_{ij} = \lambda e_{\alpha\alpha}\delta_{ij} + 2\mu e_{ij} - \xi\theta\delta_{ij} \tag{1.35}$$

where $\xi = \alpha(3\lambda + 2\mu)$; α being the coefficient thermal expansion coefficients and θ is temperature.

b) transversely isotropic material are given by[96]

$$\sigma_{rr} = C_{11}e_{rr} + (C_{11} - 2C_{66})e_{\theta\theta} + C_{13}e_{zz} - \beta_1\theta$$

$$\sigma_{\theta\theta} = (C_{11} - 2C_{66})e_{rr} + C_{11}e_{\theta\theta} + C_{13}e_{zz} - \beta_2\theta$$

$$\sigma_{zz} = C_{13}e_{rr} + C_{13}e_{\theta\theta} + C_{33}e_{zz} - \beta_2\theta$$

$$\sigma_{rz} = \sigma_{r\theta} = \sigma_{\theta z} = 0$$

where $\beta_1 = C_{11}\alpha_1 + 2C_{12}\alpha_2$, $\beta_2 = C_{12}\alpha_1 + (C_{22} + C_{33})\alpha_2$, and α_1 are the coefficient of linear thermal expansion along the axis of symmetry and α_2 is the corresponding quantities orthogonal to axis of symmetry, C_{ij} are the elastic material parameters (constants) and θ is the temperature change.

For the special case of steady heat flow, we have

$$\nabla^2\theta = 0 \tag{1.36}$$

1.1.4.2 PLASTIC STATE

Metals obey Hooke's law only in a range of small strain. When a metal is strained beyond an elastic limit, the law no longer applies. The behavior of metals beyond their elastic limit is complicated. For the analysis of continuum stress and strain distributions, a constitutive theory of plasticity should satisfy the yield condition under combined stresses.

1.1.4.2.1 Yield Criterion

Yield criterion is a condition that defines the limit of elasticity and the beginning of plastic deformation under any possible combination of stresses. It is a mathematical expression of the stress states that will cause yielding or plastic flow.

Mathematically,

$$f(\sigma)\begin{cases} = k, & \text{elastic-plastic behavior} \\ < k \text{ or } f(\sigma) = k \wedge \dot{f}(\sigma) < 0, & \text{elastic behavior} \\ > k, & \text{has no physical meaning.} \end{cases} \quad (1.37)$$

where f is a yield function, σ is the stress tensor, k is a constant (or critical value) which can be determined experimentally, $\dot{f}$ is a material time derivative.

It is convenient to express yield criteria in terms of principal stresses σ_{11}, σ_{22}, and σ_{33} for elastic-plastic behavior.

Most commonly used yield criterion in classical treatment of problems in elastic-plastic behavior.

1.1.4.2.2 *Isotropic Yield Criteria*

The three most commonly used yield conditions for isotropic metallic materials are:

a) Tresca yield criteria
   One of the earliest yield criteria is proposed by Tresca. He assumed that plastic yield occurs when a critical value k of the shear stress is reached:

$$f = \max\left\{ \frac{1}{2}|\sigma_{22} - \sigma_{33}|, \ \frac{1}{2}|\sigma_{11} - \sigma_{33}|, \ \frac{1}{2}|\sigma_{22} - \sigma_{11}| \right\}$$

b) von Mises yield criteria
   von Mises proposed a criterion which states that plastic yielding occurs when the following condition is satisfied:

$$f = \frac{(\sigma_{22} - \sigma_{33})^2 + (\sigma_{11} - \sigma_{33})^2 + (\sigma_{11} - \sigma_{22})^2}{2} = 3k^2 \quad (1.38)$$

This criterion gives better predictions than the Tresca criterion. Moreover, it is more suitable in numerical models, since sharp corners in the yield surface are absent.

Effective stress $\bar{\sigma}$ defined as

$$\bar{\sigma} = \sqrt{f} = \sqrt{\frac{\left(\sigma_{22} - \sigma_{33}\right)^2 + \left(\sigma_{11} - \sigma_{33}\right)^2 + \left(\sigma_{11} - \sigma_{22}\right)^2}{2}} \qquad (1.39)$$

c) Hoffman yield criteria

Hoffman proposed a criterion which states that plastic yielding occurs when the following condition is satisfied:

$$f = \frac{\left(\sigma_{22} - \sigma_{33}\right)^2 + \left(\sigma_{11} - \sigma_{33}\right)^2 + \left(\sigma_{11} - \sigma_{22}\right)^2}{2} +$$

$$\left(f_c - f_t\right)\left(\sigma_{11} + \sigma_{22} + \sigma_{33}\right) = f_c f_t \qquad (1.40)$$

where f_c and f_t are the uniaxial compressive and tensile yield stresses, respectively.

Effective stress $\bar{\sigma}$ defined as

$$\bar{\sigma} = \sqrt{f} \qquad (1.41)$$

1.1.4.2.3   *Anisotropic Yield Criteria*

The most widely used criterion for anisotropic materials are:

a) Hill yield criteria

Hill states that plastic yielding occurs when the following condition is satisfied:

$$2f = F\left(\sigma_{22} - \sigma_{33}\right)^2 + G\left(\sigma_{11} - \sigma_{33}\right)^2 + H\left(\sigma_{11} - \sigma_{22}\right)^2 = 1 \qquad (1.42)$$

where F, G and H are parameters that can be computed from experimentally determined normal yield stresses.

Effective stress $\bar{\sigma}$ defined as

$$\bar{\sigma} = \frac{1}{\sqrt{G + H}} \sqrt{2f} \qquad (1.43)$$

b) Singh and Ray yield criteria

Singh and Ray proposed a criterion which states that plastic yielding occurs when the following condition is satisfied:

$$2f = F\left(\sigma_{22} - \sigma_{33}\right)^2 + G\left(\sigma_{11} - \sigma_{33}\right)^2 + H\left(\sigma_{11} - \sigma_{22}\right)^2$$

$$+ \left(k_1\sigma_{11} + k_2\sigma_{22} + k_3\sigma_{33}\right) = 1 \qquad (1.44)$$

where F, G, and H are parameters in Hill and $k_i = \left[\left(f_c - f_t \right) / f_c f_t \right]_i$, $i=1,2,3$, involve uniaxial compressive and tensile stresses of in principal directions 1, 2, 3, respectively.

Effective stress $\bar{\sigma}$ defined as

$$\bar{\sigma} = \frac{1}{\sqrt{G+H}} \sqrt{2f} \tag{1.45}$$

When the deformation reaches a certain limit, the elastic limit, the material starts to deform plastically, and the strain deviation e'_{ij} is no longer given by Hooke's law. In this case, we define the plastic strain increment $de_{ij}^{(p)}$ as the actual strain deviation increment de'_{ij} minus the elastic strain increment $de_{ij}^{'(e)}$ computed from Hooke's law as if it would be still applied.

For an ideal plastic solid that obeys von Mises'[120] yield criterion and flow rule, the following simplifications are considered:

a) The plastic strain assumes incompressibility and the plastic strain deviation tensor is same as the plastic strain tensor.
b) The material is in elastic state and obeys Hooke's law, that is, $\dot{e}_{ij}^p = 0$ as long as the second invariant $J_2 < k^2$, where k is a constant.
c) Yielding takes place (or elastic limit is reached) only when $J_2 < k^2$. Then the rate of change of plastic strain is proportional to the stress deviation, $\dot{e}_{ij}^p = \sigma'/\mu$, $\mu > 0$ where μ is a proportional factor.
d) Any stress state corresponding to $J_2 > k^2$ cannot be realized in the material.

In the classical treatments, the ideal theories of elasticity and plasticity are treated separately and then combined together through a semi-empirical law, called the yield condition. What actually happens is that when a medium starts to yield, a constraint is placed on the invariant of the strain tensor of the field such that they satisfy functional relation of the form

$$f\left(I_1, I_2, I_3\right) = 0 \tag{1.46}$$

where I_1, I_2, and I_3 are the first, second, and third strain invariant.

The form of f should be determined from the condition that the modulus of transformation takes on a singular value like zero or infinity and not by any ad-hoc conditions. If we assume the incompressibility of the material and I_3 is very small, thus equation (1.46) can be reduced to the form

$$g\left(I_2\right) = \text{a constant} \tag{1.47}$$

which is known as the Huber–von Mises' yield condition. If two of the principal stresses are equal or one is the arithmetic mean of the other two conditions[47] reduces to the Tresca's yield condition. But it is clear that equation[47] cannot account for the Bauschinger effect, for which I_1 must appear in the yield condition. Seth[93] has expressed that when the material is in the fully plastic state then the yield stress Y in simple tension is $Y = E / n$ where E the response coefficient in the transition range. It is also concluded that the yield stress in compression is twice than in tension and the general form of yield condition also contains the Bauschinger effect.

The stress–strain relationships for an elastic-plastic solid were first proposed by Prandtl[77] for the case of plane strain deformation. The general form of the equations was given by Reuss.[81] Reuss assumed that increment in plastic strain, denoted by a superscript "p" in the following equations, is at any instant proportional to the instantaneous stress deviation σ'_{ij} and shear stresses. Thus,

$$\frac{de^p_{11}}{\sigma'_{11}} = \frac{de^p_{22}}{\sigma'_{22}} = \frac{de^p_{33}}{\sigma'_{33}} = \frac{de^p_{12}}{\sigma'_{12}} = \frac{de^p_{23}}{\sigma'_{23}} = \frac{de^p_{31}}{\sigma'_{31}} = d\lambda$$

or   (1.48)

$$\frac{de^p_{ij}}{\sigma'_{ij}} = d\lambda, \quad (i, j = 1, 2, 3)$$

where λ is an instantaneous positive constant of proportionality which may vary throughout a straining program.

$$de^e_{11} = \frac{1}{E}(d\sigma_{11} - v(d\sigma_{22} + d\sigma_{33})$$

$$de^e_{11} = \frac{1+v}{E}(d\sigma_{11} - d\sigma_m) + \frac{(1-2v)}{E}d\sigma_m$$

$$de^e_{11} = \frac{1}{2G}d\sigma'_{11} + \frac{(1-2v)}{E}d\sigma_m$$

where $\sigma'_{11} = \sigma_{11} - \sigma_m$ is deviatoric component, $\sigma_m = (1/3)(\sigma_{11} + \sigma_{22} + \sigma_{33})$ is hydrostatic stress and $G = E/2(1+v)$ is the torsion modulus.

$$2Ge^e_{xy} = \sigma_{xy}$$

Equations (1.48) do not give direct information about their absolute magnitude. The total increment in strain is equal to the sum of the elastic strain increment (denoted by a superscript "*e*") and the plastic strain increment. Thus,

$$de_{ij} = de_{ij}^p + de_{ij}^e = \sigma'_{ij} d\lambda + \left\{ \frac{1}{2G} d\sigma'_{ij} + \frac{(1-2\nu)}{E} \delta_{ij} d\sigma_m \right\} \qquad (1.49)$$

where ν is the Poisson ratio. Since the plastic strain causes no change in plastic volume, then condition of incompressibility can be written in terms of the principal or normal strains as

$$de_{11}^p + de_{22}^p + de_{33}^p = 0 \quad \text{or} \quad de_{ii}^p = 0 \qquad (1.50)$$

Equation (1.48) then gives

$$\frac{de_{11}^p - de_{22}^p}{\sigma'_{11} - \sigma'_{22}} = \frac{de_{22}^p - de_{33}^p}{\sigma'_{22} - \sigma'_{33}} = \frac{de_{33}^p - de_{11}^p}{\sigma'_{33} - \sigma'_{11}} = d\lambda \qquad (1.51)$$

Equation (1.51) states that Mohr circles of stress and plastic strain increment are similar.

Equation (1.48) can be rewritten in terms of normal stresses and resulting equations are

$$de_{11}^p = (d\lambda)\sigma'_{11} = (d\lambda)\frac{2}{3}\left[\sigma_{11} - \frac{1}{2}(\sigma_{22} - \sigma_{33})\right] \qquad (1.52)$$

Since $\sigma'_{ij} = \sigma_{ij} - \frac{1}{3}\delta_{ij}\sigma_{kk}; \quad i, j, k = 1, 2, 3$

Equation (1.49) thus consists of three equations of the type

$$de_{11} = \frac{2}{3}d\lambda\left[\sigma_{11} - \frac{1}{2}(\sigma_{22} - \sigma_{33})\right] + \frac{d\sigma_{11} - \nu(d\sigma_{22} + d\sigma_{33})}{E} \qquad (1.53)$$

and three equation of the type

$$de_{23} = \sigma_{23}d\lambda + \frac{d\sigma_{23}}{2G} \qquad (1.54)$$

Finally from equation (1.49), it is observed that the volumetric and deviatoric strain increments are separated in the expression of total strain increment. Including the von Mises' yield criterion, the Prandtl–Reuss equations are also written as

$$de'_{ij} = \sigma'_{ij}d\lambda + \frac{1}{2G}d\sigma'_{ij}, \quad de_{ij} = \frac{(1-2\nu)}{E}d\sigma_{ii}, \quad \sigma'_{ij}\sigma'_{ij} = 2k^2 \qquad (1.55)$$

These equations for elastic-plastic solid are generally difficult to handle in case of real problems and, as a result of this very few solutions are available. In problems of large deformations, the elastic strains may often be neglected altogether. The material is then considered as perfectly plastic solid. When the stresses are below the yield point of the material, no straining takes place, and the total increments in strain are identical. Stress–strain relations for such type of materials are proposed by Levy and von Mises. In presenting the relationship between stress and strain, we have not followed the historical development of the field. At present time, it is more logical to consider the Levy–von Mises equations as special form of the Prandtl–Reuss equations. However, it was first proposed by Saint-Venant (1870) that principal axes of strain increment coincided with the axes of principal stress. The general relationship between strain increment and the reduced stresses was first introduced by Levy (1871) and independently by von Mises (1913). These equations are now known as Levy–von Mises equations and are written as

$$\frac{de_{11}}{\sigma'_{11}} = \frac{de_{22}}{\sigma'_{22}} = \frac{de_{33}}{\sigma'_{33}} = \frac{de_{12}}{\sigma'_{12}} = \frac{de_{23}}{\sigma'_{23}} = \frac{de_{31}}{\sigma'_{31}} = d\lambda \qquad (1.56)$$

The superscript "p" of equation (1.56) may be dropped, since the total strain increments are identical. Further, the Mohr circle of stress and strain increment is identical. In terms of total stresses, the Levy–von Mises relation has three equations of the type

$$de_{11} = \frac{2}{3}d\lambda\left[\sigma_{11} - \frac{1}{2}(\sigma_{22} - \sigma_{33})\right] \qquad (1.57)$$

and three equation of the type

$$de_{23} = \sigma_{23}d\lambda \qquad (1.58)$$

Since the elastic strains are not taken into account, the Levy–von Mises relations obviously cannot be used to obtain information about "Elastic Spring-back" or residual stresses. Prandtl–Reuss equations are useful in such cases.

1.1.4.3 CREEP STATE

Creep problems are complex as compared with plasticity problems. Laboratory creep tests with complex stress conditions present technical difficulties and the experiments must be performed very carefully if the results are to be reasonably reliable. Therefore, the available experiment data is slender and does not provide a reliable basis for a creep theory that is capable of describing the behavior of materials under complex stresses. Moreover, test can only be made with plane stresses and we have no information about creep performance with stresses on the three axes. Like plasticity theory, the theory of creep under complex stress is based on certain speculative considerations, which are only partially confirmed experimentally. There are many ways in which the theory can be extended to varying stressed states. In real objects, the nature of stressed state usually varies comparatively little with time, and therefore, the different theories lead to different results. For steady state of creep, Odqvist[53] has formulated the constitutive equations by considering the rate of strain energy function $\dot{w}$ with von Mises yield criterion. It relates the rate of steady state of creep to the second invariant of the stress deviator tensor in the following form

$$\dot{e}_{ij} = \frac{\partial \dot{w}}{\partial \sigma'_{ij}} = \frac{3}{2}\left(\frac{\sigma_e}{\sigma_c}\right)^{n-1}\frac{\sigma'_{ij}}{\sigma_e} \tag{1.59}$$

where $\dot{e}_{ij}, \sigma'_{ij}, \sigma_e$ are the strain rate tensor, stress deviator tensor, and effective stress, respectively; σ_c and n are material constants.

Stress–strain relations in this form are mostly used to analyze the creep problems based on the following hypothesis:

1. The material is incompressible.
2. Creep rate is independent of superimposed hydrostatic pressure.
3. Existence of flow potential with von Mises yield condition.
4. Norton's law holds in the special case, that is, for uniaxial case.

An alternate approach to the problem of multi-axial stationary creep is made possible by Wahl[122] who considered maximum shear stress (Tresca) as a stress invariant together with the associated flow rule for the body relations. In classical treatment, different constitutive equations are used for each state, based on some hypothesis, which simplifies the problem to some extent. First, the deformations are assumed to be small to make infinitesimal strain theory applicable. Second, the constitutive equations of the material are simplified by assuming incompressibility of the material and in some cases without this assumption; it is not even possible to find the solution of the problem in closed form. By using Seth's transition theory, it has been shown that the same constitutive equations are used for different states, though the elastic constants have different meanings in each state.

Equations of equilibrium in cylindrical polar coordinates for a body, having variable thickness in the radial direction are given by

$$\frac{\partial (h\sigma_{rr})}{\partial r} + \frac{h}{r}\left(\frac{\partial \sigma_{r\theta}}{\partial \theta}\right) + h\left(\frac{\partial \sigma_{rz}}{\partial z}\right) + \frac{h}{r}(\sigma_{rr} - \sigma_{\theta\theta}) + hf_r = 0,$$

$$\frac{\partial (h\sigma_{r\theta})}{\partial r} + \frac{h}{r}\left(\frac{\partial \sigma_{\theta\theta}}{\partial \theta}\right) + h\left(\frac{\partial \sigma_{\theta z}}{\partial z}\right) + 2\frac{h}{r}\sigma_{r\theta} + hf_\theta = 0, \qquad (1.60)$$

$$\frac{\partial (h\sigma_{zr})}{\partial r} + \frac{h}{r}\left(\frac{\partial \sigma_{z\theta}}{\partial \theta}\right) + h\left(\frac{\partial \sigma_{zz}}{\partial z}\right) + \frac{h}{r}\sigma_{zr} + hf_z = 0,$$

where f_r, f_θ, f_z are the body forces along r, θ, z direction, respectively.

For a body having constant thickness, equations (1.60) becomes

$$\frac{\partial (\sigma_{rr})}{\partial r} + \frac{1}{r}\left(\frac{\partial \sigma_{r\theta}}{\partial \theta}\right) + \left(\frac{\partial \sigma_{rz}}{\partial z}\right) + \frac{1}{r}(\sigma_{rr} - \sigma_{\theta\theta}) + f_r = 0,$$

$$\frac{\partial (\sigma_{r\theta})}{\partial r} + \frac{1}{r}\left(\frac{\partial \sigma_{\theta\theta}}{\partial \theta}\right) + \left(\frac{\partial \sigma_{\theta z}}{\partial z}\right) + 2\frac{1}{r}\sigma_{r\theta} + f_\theta = 0, \qquad (1.61)$$

$$\frac{\partial (\sigma_{zr})}{\partial r} + \frac{1}{r}\left(\frac{\partial \sigma_{z\theta}}{\partial \theta}\right) + \left(\frac{\partial \sigma_{zz}}{\partial z}\right) + \frac{1}{r}\sigma_{zr} + f_z = 0,$$

For rotating discs and rotating cylinders, the axis of rotation is taken along z-axis and hence the body force is the centrifugal force with components $f_r = \rho\omega^2 r$; $f_\theta = 0$; $f_z = 0$, where ρ is the density of the material of the disc.

1.1.5   COMPOSITE MATERIALS AND FUNCTIONALLY GRADED MATERIAL

Materials have always been a crucial part of humans from the time the first man was created. As time progressed, so did the enhancement of technology and knowledge. Technological advancement is associated with continuous improvement of existing material properties and the invention of new classes/types of structural material.

1.1.5.1 COMPOSITE MATERIALS

Man started to engineer their own materials from existing raw materials. These engineered materials go back to 1500 BC in the form of composites of straw and mud. The first uses of composites date back to the 1500 BC when early Egyptians and Mesopotamian settlers used a mixture of mud and straw to create strong and durable buildings. Straw continued to provide reinforcement to ancient composite products including pottery and boats. Later, in 1200 AD, the Mongols invented the first composite bow. Using a combination of wood, bone, and "animal glue," bows were pressed and wrapped with birch bark. These bows were extremely powerful and extremely accurate. Composite Mongolian bows provided Genghis Khan with military dominance, and because of the composite technology, this weapon was the most powerful weapon on earth until the invention of gunpowder.

Composite material is a material consisting of two or more chemically different constituents combined on macroscopic scale to form a useful third material, whose mechanical performance and properties are designed to be superior to those of the constituent materials acting independently. Applications of composites abound and continue to expand. They include aerospace, aircraft, automotive, marine, energy, infrastructure, armor, biomedical, and recreational (sports) applications.

1.1.5.2 FUNCTIONALLY GRADED MATERIALS

Composite materials improved over time and in the last four decades, the world was introduced with more sophisticated composites such as fiber-reinforced plastics. However, due to the limitations of delamination in such composites materials, another breed of composite material has given birth in the form of functionally graded material (FGM). FGM is not new to us. Our nature is surrounded with FGMs. For example, the bone, human skin, and the bamboo tree are all different forms of FGM. These breed of materials are the advanced materials in the family of engineering composites made of two or more constituent phases with continuous and smoothly varying composition.[35] FGMs are engineered based on different gradients of composition in the preferred material axis orientation. Due to this flexibility, FGMs are superior to homogeneous material composed of similar constituents.

FGMs are materials that are designed with changing properties over the volume of the bulk material, with the aim of performing a set of specified functions. The properties of FGMs are not uniform across the entire material, and the properties depend on the special position of the material in the bulk structure of the material.

FGMs have good potential as a substitute material where the operating conditions are severe. Examples include coatings, heat exchanger tubes, flywheels, biomedical implants, and turbine blades, no name a few. Usually, coatings are just a layer sprayed over the substrate. Overtime due to severe operating conditions and abrupt transition of material properties from the coating to the substrate, high interlaminar stresses will exist, causing the spray to be worn or peeled off from the substrate. These sudden abrupt changes can be overcome by smooth spatial grading of the material constituents.[18,36] FGM is also findings its way into new applications such as nuclear fuel pellets, plasma wall of nuclear reactor, rocket space frame components, artificial bones, dentistry, artificial skins, building materials, sport goods, thermoelectric generators, optical fibers and lenses.[50]

1.2   REVIEW OF LITERATURE

1.2.1   ANALYSIS OF ROTATING DISCS

Rotating discs are found in a range of engineering applications such as gas turbine engines, flywheels, gears, compressors, computer disc drivers, disc brakes of cars, shrink fits, circular saws, storage devices (hard disks, blue ray disks, etc.), and so on. The use of rotating discs in engineering application has generated considerable interest in many problems and has made a longstanding research area in the domain of mechanics of solid.

The problem of rotating discs was first treated in the early 19th century. As a matter of fact, the stresses in discs depend on the angular velocity with which they rotate. The analysis of stress distribution in rotating disc at high speed is of great practical importance for a better understanding of the behavior of the disc and optimum design of structures for the required purpose. The theoretical treatment of rotating disc was started by Låszlö[37] in 1925 and since then interest in this problem has never ceased.

1.2.1.1   ANALYSIS OF ELASTIC-PLASTIC TRANSITION IN ROTATING DISC

Research on elastic-plastic analysis of rotating disc is categorized into two types of treatment which are under the category of classical treatment of elastic-plastic analysis (with empirical assumption) or under the category of nonclassical treatment of elastic-plastic analysis (without empirical assumption) and hence the literature in each case will be discussed.

1.2.1.1.1   Classical Treatment of Problems

In analyzing the problem in classical treatment of elastic-plastic, researchers used some simplifying assumptions: the deformation is small enough to make infinitesimal strain theory applicable; simplifications were made regarding the constitutive equations of the material like incompressibility of the material and yield criterion. Incompressibility of the material is one of the most important assumptions, which simplifies the problem. In fact, in most of the cases, it is not possible to find a solution in closed form without this assumption.

The problems of rotating discs have been performed under various interesting assumptions and the topic can be easily found in most of the standard elasticity books. The analytical elasticity solutions of such rotating discs are available in many books of elasticity. Investigations pertaining to the behavior of rotating discs within elastic zone can be traced back to Thompson,[115] wherein he provided a numerical approach to the turbine disc. Manson[46] has presented a finite-difference solution of the equilibrium and compatibility equations for elastic stresses in a symmetrical disc, and subsequently Manson[47] has reported a simplified method for determining the disc profile under the combination of centrifugal and thermal loading. Leopold[41] has solved similar problems for discs with variable thickness using semigraphical method.

Timoshenko and Goodier[114] discussed the analysis of thin-rotating discs made of isotropic material which is found on their standard textbook. Theoretical analysis of rotating discs in plastic regime or in region of permanent deformation was first introduced by Låszlö[38] in 1948. Millenson and Manson[49] have analyzed the stress distribution in rotating disc under conditions of plastic flow and creep. Lee[40] has presented an exact solution based on deformation theory of plasticity with axial symmetry in strain hardening range and subsequently reported a partly linearized solution of plastic deformation of rotating disc considering finite strain. Manson[48] has also presented solutions for discs of work hardening materials based on von Mises theory and deformation theory of plasticity. Heyman[33] solved the problem for the plastic state by utilizing the solution in the elastic range and considering the plastic state with the help of Tresca's, von Mises or any other classical yield condition. Chakravorty and Choudhuri[7] investigated the problem of a thin rotating disc with a central circular hole and thickness varying with the radial distance by assuming two arbitrary functions for the thickness of the disc in the elasto-plastic range.

Gamer[20–22] studied the elastic-plastic linear strain hardening problem of annular disc with constant thickness under external pressure that he understood the analysis on plasticity based on Tresca's condition only is not meaningful since the corresponding displacement field is incompatible with the necessary continuity requirements at the elastic-plastic interface for rotating solid disc. Later on, Guven[30] extended Gamer's work to annular disc and solid discs of variable thickness and variable density, obtained their analytical solutions using the same material behavior, yield condition, and to fully plastic variable thickness solid discs with constant thickness in the central portion. Then researchers understood the nonlinear character of elastic-plastic rotating disc. Guven[32] considered an annular disc profile in exponential form and studied the effect of application of external pressure analytically. Guven[31] investigated the deformation of constant thickness rotating annular discs with rigid inclusion in the fully plastic state. He also obtained an analytical solution by using Tresca's yield condition and assuming linear strain hardening.

You et al.[126] proposed a polynomial stress–strain relation for nonlinear strain hardening material as most of the materials exhibit nonlinear strain hardening behavior and this nonlinearity is present in the transition region from elastic to plastic parts of stress–strain curve. Numerical method such as finite difference method is an effective technique to investigate stresses and strains for these rotating discs. However, for scientific research and engineering analysis, analytical methods and numerical methods are still very active. To determine stresses and displacement, You et al.[126] used perturbation solution technique for rotating solid disc with nonlinear strain-hardening rule. Later on, You et al.[127] studied elastic-plastic stresses in rotating discs with constant thickness and constant density using an approximate analytical method. Further, You et al.[128] obtained elastic-plastic stresses in a rotating disc whose thickness and density varying radially.

Rees[80] studied elastic-plastic deformation of rotating solid and annular uniform thickness discs made of elastic perfectly plastic material. Eraslan[12] extended the work of Rees[80] to variable thickness solid discs made of elastic linearly hardening materials and studied inelastic stress state of solid discs with exponential thickness variation using both Tresca's and von Mises criterion. Eraslan[13] studied inelastic deformations of constant and variable thickness rotating annular discs with rigid inclusion using von Mises yield criterion. Eraslan[14] obtained exact solutions to thermally induces ax-symmetric purely elastic stress distributions in nonuniform heat-generating composite tubes. Eraslan[15] also presented the analytical solution of elastic-plastic rotating annular discs with variable thickness in a parabolic form.

Eraslan and Argeso[9] calculated the elastic and plastic limit angular speed for rotating discs of variable thickness in power function form. Eraslan and Orcan[10] studied the elastic-plastic deformation of variable thickness solid discs having concave profiles. They have presented an analytical solution for elastic and plastic deformation of linearly hardening rotating solid disc of variable thickness in an exponential form. Eraslan and Orcan[11] also obtained an analytical solution for elastic-plastic deformation of a linearly hardening rotating solid disc of variable thickness in a power function form. Eraslan[16] presented an analytical solution for rotating discs with elliptical thickness variation. Apatay and Eraslan[2] have presented analytical solutions for elastic deformation of rotating solid and annular disc with parabolically varying thickness with free, radially constrained and pressurized boundary conditions. Eraslan et al.[17] studied elastoplastic deformation of variable thickness annular discs subjected to external pressure based on von Mises yield criterion, deformation theory of plasticity and Swift's hardening law.

Sharma and Sanehlata[99] analyzed the stresses for annular disc having exponential variable thickness and exponential variable density with nonlinear strain hardening material behavior under the assumption of von Mises' yield condition using finite difference method and concluded that disc whose thickness decreases radially and density increases radially is on the safer side of design as compared with the disc with exponentially varying thickness and exponentially varying density as well as to flat disc.

Sahni and Sharma[97] studied the linear strain hardening material behavior of rotating solid disc with variable density in an exponential form using Tresca's yield condition and its associated flow rule and they developed a useful analytic solution for elastic-plastic rotating solid disc with thickness and density varying in an exponential form under the assumption of Tresca's yield condition, its associated flow rule and linear strain hardening and concluded that with the variation in density exponentially (decreases radially), high angular speed is required for a material to become fully plastic which in turns give more significant and economic design by an appropriate choice of density parameters.

Most classical treatment of elastic-plastic researchers followed the assumption of either Tresca's yield condition, linear hardening and associated flow like Sahni and Sharma[97] elastic-plastic solid disc or von Mises yield condition and nonlinear strain hardening like Sharma and Sanehlata[99] using numerical methods for elastic-plastic annular disc for their solution of the problem. The general procedure for each solution of the problem is written below:

For the first case

Divide the disc into three regions, where the plastic core consists of two parts with different forms of the yield conditions, the inner being a corner regime and the outer a side regime of Tresca's hexagon.

- For the inner plastic region:
  - In this case, the stress state lies in a corner regime of Tresca's hexagon so that the radial stress is equal to the tangential stress and hence according to Tresca criterion, this stress is equal to the yield stress.
  - Then combining this stress to equilibrium equation of motion yields the stress values.
  - For a linearly strain hardening material behavior, the yield stress is given and then obtained equivalent plastic increment using associated flow rule.
  - Relate strains with radial displacement by the assumption of axis-symmetric problems for small strains.
  - Decompose the total strain into their elastic and plastic parts and using strain displacement relations one obtains the displacement.
  - Finally, the plastic strain components are obtained by subtracting their elastic parts from their total strains.

- For the outer-plastic region:
  - In this case, the stress state lies in a side regime of Tresca's hexagon so that tangential stress is greater than radial stress. Hence according to Tresca criterion, the yield stress is equal to the tangential stress.
  - Considering the equivalent plastic increment is equal to tangential plastic strain and using the associated flow rule with the yield condition gives radial plastic strain is zero and using strain hardening we can compute the tangential plastic strain.
  - Write strain in terms of stresses and equate with strain displacement equation and then simultaneously solve for tangential and radial stresses as a function of displacement.
  - Substituting the value of tangential and radial stresses into equilibrium equation of motion, we obtain the displacement. Again substituting the displacement value into the tangential and radial stresses which is a function of displacement, and finally, we obtained the plastic strain components.

- For elastic region:
  - Write the tangential and radial strains in terms of displacement.
  - Substituting the strains value in the constitutive stress–strain equation.
  - Substituting the stresses into equilibrium equation of motion and then one can obtain the displacement.
  - Finally, once we obtain the displacement, then we get strains and stresses.

For the second case

- Use infinitesimal strain displacement relation.
- Compatibility equation can be derived from the strain displacement relations.
- Use constitutive equation of elasticity to write elastic tangential and radial strain in terms of stresses. Similarly, write plastic tangential and radial strain by using Levy–von Mises equation and use total strain components.
- Define a stress function and take the assumption of tangential and radial stresses values in terms of the stress function.
- Substituting the elastic strain and the assumption value of stresses in total strain components and obtain tangential and radial strain.

- Substituting the assumed value of thickness and density and the tangential and radial strain in the compatible equation yield second-order differential equation in terms of stress function.
- Use equivalent total strain which is defined by You et al. nonlinear strain-hardening material model.
- Substitute the equivalent strain in the plastic tangential and radial strain.
- Obtain a governing differential equation by substituting the plastic tangential and radial strain into the second-order differential equation in terms of stress function.
- The second-order nonlinear differential equation under the boundary condition can be solved by using different numerical methods.

1.2.1.1.2   Nonclassical Treatment of Problems

In analyzing the problem in nonclassical treatment of elastic-plastic transitional stresses in a rotating disc, researchers used Seth's transition and generalized strain measure theory that does not require some simplifying empirical assumptions like yield criterion and the associated flow rule. The asymptotic solution through the critical points of differential system defining the deformed field and has successfully been applied to a more general and large number of problems.

Suresh[112] analyzed thermo-elastic-plastic transition in rotating discs with steady-state temperature by using Seth's transition theory. The results obtained here are applicable to compressible materials. He derived expression for stresses in case of thermo-elastic-plastic transitions in rotating discs with steady-state temperature. It is observed that if the additional condition of incompressibility is imposed, then the expressions for stresses correspond to those arising from Tresca criterion.

Gupta and Shukla[23] analyzed the effect of nonhomogeneity on elastic-plastic transition in a thin annular rotating disc by Seth's transition theory. It is observed that in the presence of nonhomogeneity in thin-rotating discs required higher angular velocity for initial yielding as compare to homogeneous disc but less percent increase in angular velocity to become fully plastic against initial yielding and this percentage goes on decreasing with the increase in nonhomogeneity.

Gupta and Kumari[26] studied elastic-plastic transition in a transversely isotropic disc with variable thickness under internal pressure. Elastic-plastic stresses derived for transversely isotropic disc having variable thickness under internal pressure using Seth's transition theory. Yielding occurred at the internal or external surface of the disc depending upon thickness ratio. The disc having variable thickness and made of transversely isotropic material yields at a less pressure as compared with disc made of isotropic material. Transversely isotropic disc requires high percentage increase in pressure to become fully plastic from initial yielding as compared with isotropic material.

Gupta and Pankaj[27] derived the elastic-plastic transition and fully plastic stresses for a thin-rotating disc with shaft at different temperatures. It is observed that the rotating disc with inclusion and made of compressible material requires lesser angular speed to yield at the internal surface and higher percentage increase in angular speed to become fully plastic as compared with disc made of incompressible material. With the introduction of thermal effect the rotating disc with inclusion required lesser angular speed to yield at the internal surface. Rotating disc made of compressible material with inclusion requires higher percentage increase in angular speed to become fully plastic as compared with disc made of incompressible material. Thermal effect also increases the values of radial and circumferential stresses at the internal surface for fully plastic state.

Pankaj and Bansal[55] derived stresses for the elastic-plastic transition and fully plastic state have been derived for a thin-rotating disc with inclusion. It is observed that the rotating disc with inclusion and made of compressible material requires lesser angular speed to yield at the internal surface, whereas it requires higher percentage increase in angular speed to become fully plastic as compared with disc made of incompressible material.

Sanjeev and Sahni[83] studied elastic-plastic transition of transversely isotropic thin-rotating disc by using Seth's transition theory. It is observed that thin-rotating disc made of transversely isotropic material yields at a higher angular speed as compared with disc made of isotropic material. Rotating disc made of isotropic material required high percentage increase in angular speed to become fully plastic from its initial yielding as compared with disc made of transversely isotropic material. Rotating disc made of transversely isotropic material is on the safer side of design as compared with rotating disc made of isotropic material.

Pankaj[56] studied elastic-plastic transition stresses in an isotropic disc having variable thickness subjected to internal pressure. They derived elastic-plastic transitional stresses in an isotropic disc having variable thickness under internal pressure derived by using Seth's transition theory. It is observed that disc made of compressible material and having variable thickness yields at some radius R1 at a higher pressure as compared with disc made of incompressible material which yields at the outer surface. Flat disc made of incompressible material yields at internal surface at higher pressure as compared with disc made of compressible material. Circumferential stress is maximum at the outer surface of the disc having variable thickness.

Pankaj[58] studied elastic-plastic transition stresses in a disc having variable thickness and Poisson's ratio subjected to internal pressure. Elastic-plastic transitional stresses in an annular disc having variable thickness and variable Poisson's ratio subjected to internal pressure has been derived by using Seth's transition theory. It is seen that thickness and Poisson's ratio variation influence significantly the stresses and pressure required for initial yielding. The thickness variation reduces the magnitude of the stress and pressure needed for fully plastic state. It is seen for fully plastic state that circumferential stresses are maximum at the outer surface.

Pankaj[59] studied elastic-plastic transition stresses in a thin-rotating disc with rigid inclusion by infinitesimal deformation under steady-state temperature. Stresses derived for the elastic-plastic transition and fully plastic state for a thin-rotating disc with rigid shaft at different temperatures. It is observed that at room temperature rotating disc made of compressible material and of smaller radii ratio yields at the internal surface at a higher angular speed as compared with rotating disc made of incompressible material. With the introduction of thermal effect the rotating disc with inclusion required lesser angular speed to yield at the internal surface. Rotating disc made of compressible material with inclusion requires higher percentage increase in angular speed to become fully-plastic as compared to disc made of incompressible material. Thermal effect also increases the values of radial and circumferential stresses at the internal surface for fully-plastic state.

Pankaj[60] derived the stresses for the elastic-plastic transition and fully plastic state for a thin-rotating disc with rigid shaft having variable thickness by using Seth's transition. It is observed that in the absence of thickness, rotating disc with inclusion and made of compressible material, for example, copper, brass, and steel, yields at the internal surface at a lesser angular speed as compared with a rotating disc made of incompressible material, for example, rubber whereas it requires a higher percentage increase in angular speed to become fully plastic. With the effect of variation thickness, higher angular speed is required to yield at the internal surface. It is observed that the radial stress is maximum at the internal surface. With the effect of variable thickness, it increases the value of radial and circumferential stress at the internal surface for transitional state, whereas rotating disc having variable thickness increases the values of radial and circumferential stress at the internal surface for fully plastic state.

Pankaj, Singh, and Jatinder[61] studied the analysis of transitional stresses in thin-rotating disc for different materials by using Seth's transition theory. It is observed that the rotating disc made of incompressible material required higher angular speed for initial yielding as compared with disc made of compressible materials. Circumferential stresses is maximum at the internal surface for compressible materials as compared with incompressible material, that is, Steel, brass, and lead materials required maximum circumferential stresses as compare to Rubber material. Rotating disc is likely to fracture by cleavage close to the bore.

Pankaj, Singh, and Jatinder[63] studied in elastic-plastic stresses in a thin-rotating disc with shaft having density variation parameter under steady-state temperature. They derived steady thermal stresses in a rotating disc with shaft having density variation parameter subjected to thermal load by using Seth's transition theory. It is observed that compressible material required higher percentage increased angular speed to become fully plastic as compared with rotating disc made of incompressible material. Circumferential stresses are maximal at the outer surface of the rotating disc. With the introduction of thermal effect, it decreases the value of radial and circumferential stresses at inner and outer surface for fully plastic state.

Pankaj, Singh, and Jatinder[62] applied Seth's transition to the problems of thickness variation parameter in a thin-rotating disc by finite deformation. The results obtained here are applicable to compressible materials. If the additional condition of incompressibility is imposed, then the expression for stresses corresponds to those arising from Tresca yield condition. It is observed that effect of thickness for incompressible material of the rotating disc required higher percentage increased in angular speed to become fully plastic as compared with rotating disc made of compressible materials. For at disc compressible materials required higher percentage increased in angular speed to become fully plastic as compared with disc made of incompressible material. With effect of thickness circumferential stresses are maximum at the external surface for compressible materials as compared with incompressible materials whereas for flats disc circumferential stresses are maximum at the internal surface for incompressible material as compared with compressible materials.

Pankaj and Singh (2015)[64] studied elastic-plastic transitional stresses distribution and displacement for transversely isotropic circular disc with inclusion subject to mechanical load and obtained rotating disc made of isotropic material required higher angular speed to yield at the internal surface as compared with disc made of transversely isotropic materials. Effect of mechanical load in a rotating disc with inclusion made of isotropic material as well as transversely isotropic materials increase the values of angular speed yield at the internal surface. With the introduction of mechanical load rotating disc made of Beryl material required maximum radial stress as compared to disc made of Mg and brass materials at the internal surface.

Pankaj, Jatinder, and Singh[65] investigated stresses and displacement in thin nonhomogeneous rotating disc using Seth's transition theory. Nonhomogeneity is assumed due to the variation of modulus of rigidity. As a numerical example, it is observed that in the presence of nonhomogeneity having values $k > 0$ at the bore, reduces stresses, displacement and the angular speed as compared with lesser value, that is, $k < 0$ of nonhomogeneity. Radial stresses maximum at the internal surface.

Jatinder, Pankaj, and Singh[34] studied steady thermal stresses in a thin-rotating disc of finitesimal deformation with edge loading by Seth's transition theory to the problems of thickness variation parameter. It is observed that for rotating disc made of compressible material required higher angular speed to yield at the internal surface as compared with disc made of incompressible material and a much higher angular speed is required to yield with the increase in radii ratio. With the introduction of thermal effects, lesser angular speed is required to yield at the internal surface. Thermal effect in the disc increase the value of circumferential stress at the internal surface and radial stresses at the external surface for compressible as compared with incompressible material.

Pankaj, Singh, and Sandeep[66] studied stress evaluation in a transversely isotropic circular disc with an inclusion. It is observed that a rotating disc made of transversely isotropic material requires a low percentage increase in angular speed to become fully plastic as compared with a rotating disc made of isotropic material. Circumferential stresses are maximum at the external surface for the fully plastic state.

Pankaj and Sandeep[73] studied the analysis of stresses in a transversely isotropic thin-rotating disc with rigid inclusion having variable density parameter by using Seth's transition theory. With the introduction of density variation parameter, lesser values of angular speed are required to yield at the internal surface for rotating disc made of isotropic/transversely isotropic materials. The radial stress is maximum at the internal surface for both isotropic and transversely isotropic materials. With the effect of density variation, parameter increases the values of stresses and displacement of the rotating disc made of isotropic/transversely isotropic materials. Radial stresses are maximum at the internal surface for both isotropic and transversely isotropic materials.

Pankaj, Sethi, Shivdev, Singh, and Emmanuel[71] presented exact solution of rotating disc with shaft problem in the elastoplastic state of stress having variable density and thickness by using Seth's transition theory. It is observed that the rotating disc made of the compressible material with an inclusion requires higher angular speed to yield at the internal surface as compared with the disc made of incompressible material, and a much higher angular speed is required to yield with the increase in radii ratio. The thickness and density parameters decrease the value of angular speed at the internal surface of the rotating disc of compressible as well as incompressible materials.

1.2.1.2   ANALYSIS OF CREEP TRANSITION IN ROTATING DISC

In many applications, rotating discs are exposed to elevated temperatures where creep deformation becomes important. A lot of progress has been made theoretically as well as experimentally to define certain aspects of the involved mechanisms of creep and in general research on creep analysis of rotating disc are categorized into two types of treatment which are under the category of classical treatment of creep analysis (with empirical assumption) or under the category of nonclassical treatment of creep analysis (without empirical assumption) and hence the literature in each case will be discussed.

1.2.1.2.1   *Classical Treatment of Problems*

In literature, the first recorded experiments relating to creep in connection with suspension bridges, measuring instruments and steam engines have been taken in the 1830s. After a long time, Andrade[1] gave the concepts of primary, secondary, and tertiary creep in case of uniaxial creep tests with constant load or stress. In 1922, Dickenson published his researches on the creep resistance of structural steel and alloy steel members in a furnace and he has given the recognition of the importance of creep to the industrial. In 1929, Norton discovered the exponential law, which applies to many metals, that is,

$$\dot{\varepsilon} = \frac{d\varepsilon}{dt} = k\bar{\sigma}^{n} \qquad (1.62)$$

where $\dot{\varepsilon}$ is strain rate, $\bar{\sigma}$ is effective stress and k and n are constants.

In around 1930s, Bailey showed that creep (time-dependent plasticity) deformation of structural metals takes place under constant volume and a superimposed hydrostatic pressure does not influence creep deformation. From these facts and the assumption of isotropy, Odqvist (1934) deduced constitutive relations for secondary creep under triaxial stresses, which have the same form as von Mises equation for time-dependent plasticity.

Wahl et al.[121] analyzed steady-state creep deformations and stress distributions in a rotating forged disc made of 12% chromium steel at 1000°F. The results obtained were validated experimentally. Theoretical analysis of creep was carried out using von Mises and Tresca yield criteria, while the creep behavior was described by power law. The stress distributions obtained using Tresca and von Mises criteria do not differ significantly. The study also reveals that the theoretical and experimental stresses are in better agreement if the Tresca criterion is used. The creep deformations estimated using von Mises theory are found to be quite low as compared with those obtained experimentally, which may be attributed to anisotropy of the disc material.

Wahl[122] derived some formulas for calculating stress distributions in rotating discs having constant and variable thickness, undergoing steady-state creep at elevated temperature. The formulas were based on the Tresca's yield criterion and the associated flow rule and give reasonable results when compared with the available experimental data.[121] The method proposed was also applied to calculate the transient change in stress when the stress distribution changes from an initial to a steady-state condition during the starting period.

Wahl[123] used the formulas derived in their previous work[121] to construct the design charts of stress distribution in constant thickness discs, undergoing steady-state creep, for different values of stress exponent (n) and diameter ratios using the formulas derived earlier. In all the cases, the discs were subjected to a radial peripheral load to simulate the effect of blade loading. The steady-state creep rate was expressed as a product of power function of stress multiplied by a function of time.

Wahl[124] extended his previous work[123] on rotating discs having central holes and undergoing steady-state creep to cases where the radial and tangential stresses are equal over all or a portion of the disc. Both constant and variable thickness discs were considered and the charts were presented for determining the ratios of peak stress to average stress for various diameter ratios, disc contours, and stress exponent. The study indicates that the variable thickness disc has somewhat lower ratios of peak to average stress than those observed for constant thickness disc.

Ma[42] studied a mathematical approach based on maximum shear-stress theory (Tresca yield criterion) associated with the von Mises flow rule with the assumption that the tangential stresses in the disc are invariably greater than the radial stresses, except at the center of the disc to calculate creep deformations and stress distributions in rotating solid discs of variable thickness operating at uniform temperature. The results obtained using von Mises criterion is found to be in excellent agreement with the available experimental creep data.[121]

Ma[43] extended his work for variable thickness solid discs, used in gas turbine and jet engine, operating at uniform temperature. The study used Tresca's criterion and its associated flow rule while the steady-state condition was described by exponential creep law. It is revealed that the stress distributions over central portion of variable thickness disc are quite different from those observed in a constant thickness disc.

Ma[44,45] further extended his analysis for variable thickness disc operating under variable temperature. The steady-state creep was described by either exponential creep law[44] or power law.[45] The study reveals that the proposed analyses can be used to obtain closed-form solutions for complex disc design problem with great simplicity instead of using tedious numerical solutions.

Wahl[125] investigated the effects of initial transient period on long-time creep tests of rotating discs by using both time-hardening and strain-hardening relations. The results obtained were applied to the long time spin-tests conducted on steel discs at 1000°F.[121] The study reveals that by considering the effects of transient period, there is no appreciable impact on the overall creep deformation noticed during the spin tests. However, when the creep deformations are of the order of elastic strains it is necessary to include such transient effects.

Arya and Bhatnagar[3] analyzed the creep stresses and deformations in rotating disc made of orthotropic materials by assuming the creep rate to be a function of time. A numerical example was worked out to investigate the effect of transition (nonsteady) creep and orthotropicity of material on stress and strain distributions in the disc. It is observed that the tangential stress at any radius and the tangential strain at the inner radius of the disc decrease at all times for an anisotropic material. The time taken to reach steady-state distribution decreases with increasing anisotropy of the disc material.

Bhatnagar et al.[4] carried out analysis of steady-state creep in rotating discs having constant and variable (linear and hyperbolic) thickness. The creep stresses and strains were obtained for different cases of anisotropy by using Norton's power law creep model. The study reveals that the selection of a certain type of material anisotropy and an optimum disc profile would lead to a better disc design.

Singh et al.[101] performed creep analysis in an anisotropic 6061Al–SiCw disc rotating at 15,000 rpm and undergoing steady-state creep described by Norton's power law at 561 K. The study revealed that the presence of anisotropy leads to significant change in the strain rates.

Singh[102] studied the effect of parameters, such as particle size, particle content, and temperature on creep response in a rotating isotropic disc made of Al–SiCp composite by using Norton's law and von Mises yields criterion. It was concluded that the effect of parameters has a significant change on tangential stress distribution, but their effect on radial stress is relatively very small. However, by decreasing particle size and the operating temperature while increasing the particle content, the strain rates in both radial and tangential directions in the disc get reduced by several orders.

The creep analysis in an isotropic functionally graded material rotating disc of Al–SiCp composite was first introduced by Singh and Ray in 2001.[103] They investigated the creep analysis in isotropic functionally graded materials rotating disc made of a composite containing silicon carbide particles in a matrix of pure aluminum at uniform elevated temperature by using Norton's power law and they assumed the particle distribution is linearly decreasing from the inner to the outer-radius along the radial distance in the disc estimated by regression t of the available experimental data. It is observed that the steady-state creep response of the functionally graded material disc in terms of strain rates is significantly superior as compared with a similar disk having uniform distribution of SiCp.

Ray and Singh[104] extended the work of Singh and Ray[103] to the analytical treatment of anisotropy and creep in orthotropic aluminum–silicon carbide composite rotating disc made of composites containing SiC whiskers under steady state using Hill yield criterion and compared with the results obtained using von Mises yield criterion for the isotropic composites. It is observed that the tangential stress distribution is lower in the middle of the disc but higher near the inner and the outer-radius but the radial stress distribution does not get significantly affected due to anisotropy and also observed that anisotropy helped to reduce the tangential strain rate significantly, more near inner radius and the strain rate distribution in the orthotropic disc is lower than that in the isotropic disc following von Mises criterion. It should be noted that the anisotropy constants taken from the experimental results of other studies and the lowering of tangential creep rate may be significant in the context of real-life engineering. The compressive radial strain rate also reduced in the disc following Hill criterion of yield plasticity as compared with that in isotropic disc. Thus, anisotropy appeared to help in restraining creep response both in the tangential and in the radial directions. Further Singh and Ray[106] extended to investigate for the effect of anisotropy on the creep behavior of rotating disc made of functionally graded Al–SiCp. The study reveals that the presence of anisotropy in the disc leads to significant reduction in the tangential and radial strain rates over the entire disc radius.

Singh and Ray[105] suggested a new yield criterion for residual stress and steady-state creep in an anisotropic composite rotating disc to describe plasticity of composite materials containing whiskers and short fibers, which in fact reduces to Hill anisotropic and Hoffman isotropic yield criterion at appropriate limits. They carried out analysis of steady-state creep in a rotating disc made of composites containing silicon carbide whiskers (Al–SiCw) using newly suggested yield criterion and compared with the results obtained using Hill anisotropic yield criterion ignoring difference in yield stresses. It is observed that the presence of residual stress in an anisotropic disc results in significant changes in the distribution of tangential stress but similar distribution of radial stress when compared with those obtained in a similar anisotropic disc but without residual stress. The presence of tensile residual stress in the disc leads to significant increase in creep rate as compared with that observed in a similar anisotropic disc without residual stress. In the presence of residual stress, the radial strain rate becomes tensile in the middle of the disc, however, it remains compressive toward the inner and outer-radii. They concluded that anisotropy and residual stress have important engineering consequences.

Singh and Ray[107] investigated the effect of thermal residual stress on the steady-state creep behavior of a rotating disc made of 6061Al–20 vol% SiCw composite using isotropic Hoffman yield criterion while describing the creep by Norton's power law. It is observed that the tensile residual stress significantly affects the strain rates in the disc as compared with the strain rate in the disc without residual stress.

Singh[108] studied the creep behavior of a whisker-reinforced aniso-tropic rotating disc made of Al–SiCw composite. He developed a one-parameter model to study the effect of anisotropy on the creep behavior of the disc by using Norton's power law with varying extent of anisotropy, characterized by a parameter α. It is observed that the radial strain rate, which remains compressive for isotropic composite disk ($\alpha = 1: 0$) and anisotropic disc having $\alpha > 1$, becomes tensile in the middle of the disc when the extent of anisotropy parameter $\alpha < 1$. By changing the extent of anisotropy from $\alpha > 1$ to $\alpha < 1$, the variation of tangential strain rate in the disc remains similar; however, its magnitude reduces by about five orders. Thus, the presence of anisotropy introduced significant change in the strain rates, though its effect on the resulting stress distribution is relatively small.

Gupta, Ray, and Singh[29] studied the role of reinforcement geometry on the steady-state creep behavior of a rotating composite disc. They developed a mathematical model to predict steady-state creep response of a rotating disc made of SiC (particle/whisker)-reinforced 6061Al matrix composite. The model investigated the effect of SiC morphology on the creep behavior of composite disc under the steady-state creep behavior described by Sherby's creep law. The creep stresses and creep rates signifi-cantly affected by the morphology of SiC. The steady-state creep rates in whisker-reinforced disc observed to be significantly lower than those observed in particle-reinforced disc.

Singh and Rattan[109] studied creep analysis of an isotropic rotating Al–SiC composite disc taking into account the phase specific. They analyzed steady-state creep in a rotating disc made of Al–SiCp composite out using isotropic Hoffman yield criterion and the results obtained are compared with those using von Mises yield criterion ignoring difference in yield stresses. It is observed that the stress distribution is not too much affected due to presence of phase specific thermal residual stress. The presence of residual stress increase the tangential strain rate particularly in the region near the outer-radius of the disc, as compared with the tangential strain rate in the disc without residual stress. The radial strain rate, which is compressive, changes significantly due to the presence of residual stress and even becomes tensile in the middle of the disc and they concluded that the presence of residual stress significantly affects the creep in an isotropic rotating disc.

Rattan, Chamoli, and Singh[51] studied the effect of anisotropy on the creep of a rotating disc of Al–SiCp composite. They investigated the steady-state creep behavior of an anisotropic rotating disc made of aluminum–silicon carbide particulate composite using Sherby's law. The study revealed that anisotropy of the material has a significant effect on the creep behavior of the rotating disc therefore while designing the disc the effect of anisotropy should be taken care of.

Vandana and Singh[115] studied modeling anisotropy and steady-state creep in a rotating disc of Al–SiCp having varying thickness. They analyzed steady-state creep in a rotating disc made of Al–SiCp composite having variable thickness using Sherby's constitutive model. The study reveals that the anisotropy and thickness profile has a significant effect on the creep behavior of rotating disc therefore while designing the disc the effect of anisotropy should be taken care of.

Vandana and Singh[116] studied influence of anisotropy on creep in an anisotropic composite rotating disc with nonlinearly varying thickness. They investigated the effect of anisotropy in terms of a single parameter indicating strengthening or weakening in the tangential direction in composite disc with hyperbolically varying thickness introduced presumably by processing or due to alignment of dispersed reinforcements during flow of the matrix. It is observed that the radial strain rate always remained compressive for the isotropic composite as well as the anisotropic disc with a greater than unity ($a = 1.3$). However, it becomes tensile in the middle region of the disc when it is less than unity ($a = 0.7$). If a is reduced from 1.3 to 0.7, the variation of tensile strain rate in the tangential direction remains similar, but the magnitude reduced, that is, the strength in tangential direction is enhanced. This study puts forward an analytical framework for the analysis of creep stresses and creep rates in an anisotropic rotating disc with hyperbolically varying thickness.

Vandana and Singh[117] investigated creep modeling in a composite rotating disc with thickness variation in presence of residual stress. They carried out the steady-state creep response in a rotating disc made of Al–SiC (particle) composite having linearly varying thickness using isotropic/anisotropic Hoffman yield criterion and results are compared with those using von Mises yield criterion/Hill's criterion ignoring difference in yield stresses by Sherby's creep law. It is concluded that the stress and strain distributions got affected from the thermal residual stress in an isotropic/anisotropic rotating disc, although the effect of residual stress on creep behavior in an anisotropic rotating disc is observed to be lower than those observed in an isotropic disc. Thus, the presence of residual stress in composite rotating disc with varying thickness needs attention for designing a disc.

Rattan, Chamoli, Singh, and Nishi[52] studied creep behavior of anisotropic functionally graded rotating discs. They investigated the creep behavior of an anisotropic rotating disc of functionally gradient material using Hill's yield criteria by Sherby's constitutive model. It is concluded that the anisotropy of the material has a significant effect on the creep behavior of the FGM disc. It is also observed that the FGM disc shows better creep behavior than the non-FGM disc.

Vandana and Singh[118] analyzed the effect of residual stress and reinforcement geometry in an anisotrpic composite rotating disc having varying thickness. The creep analysis carried out in a rotating disc made of Al–SiC (particle/whisker) composite having hyperbolically varying thickness using anisotropic Hoffman yield criterion and results obtained are compared with those using Hill's criterion ignoring difference in yield stresses by Sherby's creep law for the steady-state creep behavior. It is observed that the stresses are not much affected by the presence of thermal residual stress, while thermal residual stress introduces significant change in the strain rates in an anisotropic rotating disc. Second, it is noticed that the steady-state creep rates in whisker-reinforced disc with/without residual stress are observed to be significantly lower than those observed in particle-reinforced disc with/without residual stress. It is concluded that the presence of residual stress in an anisotropic disc with varying thickness needs attention for designing a disc.

Sahni and Sharma[82] studied elastic-plastic rotating solid disc of exponentially varying thickness and exponentially varying density. It is concluded that high angular speed is required for a material to yield exponentially variable thickness and exponentially variable density and that becomes plastic as compared to the disc with other parameters, which in turns give more significant and economic design by an appropriate choice of thickness and density parameters.

Vandana and Singh[119] studied mathematical modeling of creep in a functionally graded rotating disc with varying thickness. This study gave an analytical framework for the analysis of creep stresses and creep rates in the isotropic rotating non-FGM/FGM disc with uniform and varying thickness by Sherby's law. The creep response of rotating disc is expressed by threshold stress with value of stress exponent as 8. The results compared for isotropic non-FGM/FGM constant thickness disc with those estimated for isotropic varying thickness disc with the same average particle content distributed uniformly and suggested the distribution of stresses and strain rates becomes relatively more uniform in the isotropic FGM hyperbolic thickness disc.

Pankaj, Nishi, and Singh[67] investigated thermal effect on the creep in a rotating disc by using Sherby's law and obtained the distributions of stresses and strain rates. They computed the creep response of a composite disc with uniform temperature for comparison with the results obtained for thermally graded discs. The creep rates in a rotating thermally graded disc can be significantly reduced in presence of thermal gradients.

1.2.1.2.2 Nonclassical Treatment of Problems

Transition theory of creep has been given by Seth in 1972 in which he has defined the measure concept in mechanics[92] and creep transition.[94] This transition theory does not require any ad-hoc assumptions like creep strain laws (or a power relationship between stress and strain). The problems of creep transitions are solved by Seth transition theory.

Suresh[111] analyzed transition theory of creep in rotating discs of variable thickness by using Seth's transition theory. It is observed that the results obtained correspond to the creep law with Tresca rule and agree with those derived by Wahl.[123]

Shukla[100] studied creep transition in a thin-rotating nonhomogeneous disc. He obtained creep stresses in a thin-rotating annular disc made of a nonhomogeneous material by using transition theory. It is observed that nonhomogeneity, with lesser value at the bore, reduces the stresses and the angular velocity required for steady state of creep as compared with the homogeneous disc, however, disc that having higher value at the bore than at the rim, increased the magnitude of angular velocity and creep stress significantly, and hence increased the possibility of a fracture in the vicinity of the bore.

Gupta et al.[24] analyzed creep stresses and strain in a thin-rotating disc having variable density by using Seth's transition theory. It is observed that a disc having radially decreasing density rotates at higher angular speed and thus has more possibility of fracture at the bore, whereas the disc with radially increasing density has lesser possibility of fracture. The creep deformation is noticed to be significant in the disc having variable density and rotating at higher angular speed. The analysis was later extended by Gupta et al.[25] to study creep stresses and strains in a thin-rotating disc having variable thickness and variable density by using Seth's transition theory. The study reveals that a rotating disc, whose density and thickness ratio decreases radially, is on the safer side of design as compared with a flat disc having variable density.

Gupta and Pankaj[28] investigated creep transition in a thin-rotating disc with rigid inclusion. It is observed that radial stress has maximum value at the internal surface of the rotating disc made of incompressible material as compared with circumferential stress and this value of radial stress further increases with the increase in angular speed. Strain rates have maximum values at the internal surface for compressible material. Rotating disc is likely to fracture by cleavage close to the inclusion at the bore.

Pankaj and Gupta[54] studied creep transition in an isotropic disc having variable thickness subjected to internal pressure. They obtained creep stresses and strain rates for an isotropic disc having variable thickness subjected to internal pressure. It is observed that circumferential stress has maximum value at the internal surface for a flat disc and at external surface for a disc having variable thickness. This value of circumferential stress further increases with the increase in measure. Strain rates have maximum values at the internal surface for at disc made of incompressible material whereas reverse is the case for the disc having variable thickness. These values of strain rates further increase with the increase in pressure and measure for a disc made of incompressible material.

Pankaj and Bansal[75] estimated creep stresses and strains in a thin-rotating disc with inclusion and having radially varying density by using Seth's transition theory. It is observed that as compared with a constant thickness disc, the disc having radially increasing density can be subjected to higher angular speed without causing fracture at the bore.

Sharma and Sahni[97] used Seth's transition theory to investigate creep stresses and strains in rotating disc made of transversely isotropic and isotropic materials. It is revealed that a transversely isotropic disc rotating at higher angular speed has more possibility of fracture at the bore as compared with isotropic disc. The possibility of fracture decreases with the increase in measure (N).

Pankaj[57] studied creep transition stresses in a thin-rotating disc with shaft by finite deformation under steady-state temperature. He derived creep stresses and strain rates for a thin-rotating disc with shaft at different temperature. It is observed that radial stress has maximum value at the internal surface of the rotating disc made of incompressible material as compared with circumferential stress and this value of radial stress further increase with the increase of angular speed. With the introduction of thermal effect, it is observed that radial stress has higher maximum value at the internal surface of the rotating disc made of incompressible material as compared with circumferential stress, and this value of radial stress further increases with the increase of angular speed as compared with the case without thermal effect. Strain rates have maximum values at the internal surface for compressible material. Rotating disc is likely to fracture by cleavage close to the inclusion at the bore.

Sanjeev and Manoj[84] obtained creep stresses for a thin-rotating disc having variable thickness and variable density with edge load. It is observed that a rotating disc whose density and thickness decreases radially with edge load is much safer for a design in comparison to a flat rotating disc having variable density. The deformation is much more significant for rotating disk with edge load than those of a rotating disc without edge load.

Pankaj, Singh, and Jatinder[62] applied Seth's transition theory to the problems of thickness variation parameter in a thin-rotating disc by finite deformation. It is observed that effect of thickness for incompressible material of the rotating disc required higher percentage increased in angular speed to become fully plastic as compared with rotating disc made of compressible materials. For flat-disc compressible materials required higher percentage increased in angular speed to become fully plastic as compared with disc made of incompressible material. With effect of thickness circumferential stresses are maximum at the external surface for compressible materials as compared with incompressible materials whereas for flats disc circumferential stresses are maximum at the internal surface for incompressible material as compared with compressible materials.

Sharma et al.[98] investigated creep stresses in a thin-rotating disc having exponentially varying thickness with inclusion and edge load. It is observed that in rotating disc made of incompressible material, the radial stress is maximum at the internal surface. Rotating disc is likely to fracture by cleavage close to inclusion at the bore, but the presence of edge load may enhance strength to the disc and avoid its fracture.

Pankaj, Jatinder, and Singh[74] obtained thermal creep stresses and strain rates for a circular rotating disc with shaft having variable density by using Seth's transition theory. It is observed that radial stress has maximum value at the internal surface of circular disc made of compressible materials as compared with tangential stress and this value of radial stress further increase with increase in angular speed. With the introduction of thermal effects, the values of radial stress further increases. Strain rates have maximum value at the internal surface for compressible materials.

Pankaj, Suresh, and Joginder[68] studied creep stresses in a rotating disc having variable density and mechanical load under steady-state temperature and obtained applicable to compressible materials. If the additional condition of incompressibility is imposed, then the expression for stresses corresponds to those arising from Tresca yield condition. With the introduction of thermal effects and load, circumferential stress, as well as radial stress, increases at the inner and outer surface of the rotating disc. The rotating disc is likely to fracture by cleavage close to the shaft at the bore.

Pankaj, Pooja, and Sandeep[69] studied creep stresses and strain rates for a transversely isotropic disc having the variable thickness under internal pressure. It is observed that the disc made of variable thickness for transversely isotropic material has a maximum circumferential stress at the outer surface in comparison to disc having constant thickness and these value further increases with the increase in measure n and k. Strain rates were maximum on the internal surface for at disc of isotropic material for measure $n = 1$ and pressure $P_i = 0.1$, but strain rates decrease at the internal surface for measure $n > 1$. The strain rate for flat disc ($k = 0$) further increases with increase in pressure at the internal surface. The disc of variable thickness ($k = 1.5$) made of isotropic material has strain rate maximum at the external surface for $P_i = 0.1$ and measure $n = 1$. These values of strain rates further increase at the external surface with the increase in pressure and variable thickness ratio, but decrease with the increase in measure n.

Pankaj, Shivdev, Nishi, and Singh[70] presented effect of mechanical load and thickness profile on creep in a rotating disc by using Seth's transition theory. It is observed that stresses increase with the increase in mechanical load and maximum value of strain rate further increases at the internal surface for compressible materials. It is concluded that rotating disc is likely to fracture by cleavage close to the shaft at the bore.

Pankaj, Sethi, Shivdev, Singh, Emmanuel, and Jasmina[72] presented modeling of creep behavior of a rotating disc in the presence of load and variable thickness by using Seth transition theory. It observed that a flat rotating disc made of compressible as well as incompressible material with load E1 = 10, increases the possibility of fracture at the bore. It is also observed that a rotating disc of incompressible material and thickness that increases radially experiences higher creep rates at the internal surface in comparison to a disc of compressible material. They suggested the model they proposed is used in mechanical and electronic devices.

1.3   FUTURE PROSPECTS

In this chapter, the analysis of elastic-plastic and creep deformation in rotating disc using classical and nonclassical treatment by different researchers has been discussed and the gap to be done are indicated. The transition and generalized strain measure theories explain a variety of physical phenomena and hence has a very wide application in all applied sciences. The study treated by nonclassical method can be extended to

- investigate elastic-plastic and creep stresses in a transversely isotropic or orthotropic rotating disc under different types of loadings such as internal pressure, external pressure, and temperature.
- investigate elastic-plastic and creep deformation in a transversely isotropic or orthotropic rotating disc made of functionally graded materials under thermal gradient.
- investigate elastic-plastic and creep response of other structural components like cylinder and shell. For instance, the analysis of elastic-plastic and creep transitional stresses and strains in an isotropic and transversely isotropic cylinder having constant or variable density under thermal gradient.
- industry that intends to create a link between the theoretical model with real structural components after performing experiments.

The solutions to the problem of elastic-plastic and creep transition by nonclassical treatment may also be compared with the solutions by classical treatment.

KEYWORDS

- creep
- elastic-plastic
- rotating disc
- strain
- stress
- transition
- yielding

REFERENCES

1. Andrade, E. N. D. C. On the Viscous Flow in Metals, and Allied Phenomena. *Proc. R. Soc. Lond. A*, **1910**, *84*, 1–12.
2. Apatay, T.; Eraslan, A. N. Elastic Deformations of Rotating Parabolic Discs: An Analytical Solution. *J. Fac. Eng. Arch. Gazi. Univ.* **2003**, *18* (2), 115–135.
3. Arya, V. K.; Bhatnagar, N. S. Creep Analysis of Rotating Orthotropic Discs. *Nucl. Eng. Des.* **1979**, *55*, 323–330.
4. Bhatnagar, N. S.; Kulkarni, P. S.; Arya, V. K. Steady-State Creep of Orthotropic Rotating Discs of Variable Thickness. *Nucl. Eng. Des.* **1986**, *91* (2), 121–144.
5. Borah, B. N. *Thermo-Elastic-Plastic Transition*, PhD Thesis, Oregon State University, Oregon, USA, 1968.
6. Boresi, A. P.; Schmidt, R. J. *Advanced Mechanics of Materials*, 6th ed. John Willy and Sons., Inc.: Hoboken, NJ, 2002.
7. Chakravorty, J. G.; Choudhary, P. K. The Elastico-Plastic Problem of a Thin Rotating Disc with Variable Thickness. *Indian J. Pure appl. Math.* **1983**, *14* (1), 70–78.
8. Deiter, G. E. *Mechanical Metallurgy*; McGraw-Hill Book Company: Maryland (London), 1928.
9. Eraslan, A. N.; Argeso, H, Limit Angular Velocities of Variable Thickness Rotating Discs. *Int. J. Solids. Struct.* **2002**, *39* (12), 3109–3130.
10. Eraslan, A. N.; Orcan, Y. On the Rotating Elastic-Plastic Solid Discs of Variable Thickness Having Concave Profiles. *Int. J. Mech. Sci.* **2002**, *44* (7), 1445–1466.
11. Eraslan, A. N.; Orcan, Y. Elastic-Plastic Deformation of a Rotating Solid Disc of Exponentially Varying Thickness. *Mech. Mater.* **2002**, *34* (7), 423–432.
12. Eraslan, A. N. Inelastic Deformations of Rotating Variable Thickness Solid Discs by Tresca and von Mises Criteria. *Int. J. Comp. Eng. Sci.* **2002**, *3* (1), 89–101.
13. Eraslan, A. N. von Mises Yield Criterion and Nonlinearly Hardening Variable Thickness Rotating Annular Discs with Rigid Inclusion. *Mech. Res. Commun.* **2002**, *29* (5), 339–350.

14. Eraslan, A. N. Thermally Induced Deformations of Composite Tubes Subjected to a Nonuniform Heat Source. *J. Therm. Stresses* **2003**, *26*, 167–193.
15. Eraslan, A. N. Elastic-Plastic Deformations of Rotating Variable Thickness Annular Discs with Free, Pressurized and Radially Constrained Boundary Conditions. *Int. J. Mech. Sci.* **2003**, *45* (4), 643–667.
16. Eraslan, A. N. Stress Distributions in Elastic-Plastic Rotating Discs with Elliptical Thickness Profiles Using Tresca and von Mises Criteria. *ZAMM* **2005**, *85* (4), 252–266.
17. Eraslan, A. N.; Orcan, Y.; Guven, U. Elastoplastic Analysis of Nonlinearly Hardening Variable Thickness Annular Discs under External Pressure. *Mech. Res. Commun.* **2005**, *32* (3), 306–315.
18. Finot, M.; Suresh, S. Small and Large Deformation of Thick and Thin Film Multilayers: Effect of Layer Geometry, Plasticity and Compositional Gradients. *J. Mech. Phys. Solids* **1996**, *44*, 683–721.
19. Fung, Y. C. *Foundations of Solid Mechanics*; Prentice-Hall: Englewood Cliffs, NJ, 1965.
20. Gamer, U. Tresca's Yield Condition and the Rotating Solid Disc. *J. Appl. Mech.* **1983**, *50* (3), 676–678.
21. Gamer, U. Elastic-Plastic Deformation of the Rotating Solid Disc. *Ing. Arch.* **1984**, *54* (5), 345–354.
22. Gamer, U. The Rotating Solid Disc in the Fully Plastic State. *Forsch. Ing. Wes.* **1984**, *50* (5), 137–140.
23. Gupta, S. K.; Shukla, R. K. Effect of Nonhomogeneity on Elastic-Plastic Transition in a Thin Rotating Disc. *Indian J. Pure Appl. Math.* **1994**, *25* (10), 1089–1097.
24. Gupta, S. K.; Sharma, S.; Pathak, S. Creep Transition in a Thin Rotating Disc of Variable Density. *Defence Sci. J.* **2000**, *50*, 147–153.
25. Gupta, S. K.; Sharma, S.; Pathak, S. Creep Transition in a Thin Rotating Disc Having Variable Thickness and Variable Density. *Indian J. Pure Appl. Math.* **2000**, *31*, 1235–1248.
26. Gupta, S. K.; Kumari, P. Elastic-Plastic Transition in a Transversely Isotropic Disc with Variable Thickness under Internal Pressure. *Indian J. Pure Appl. Math.* **2005**, *36* (6), 329–344.
27. Gupta, S. K.; Pankaj, T. Thermo-Elastic-Plastic Transition in a Thin Rotating Disc with Inclusion. *Therm. Sci.* **2007**, *11* (1), 103–118.
28. Gupta, S. K.; Pankaj, T. Creep Transition in a Thin Rotating Disc with Rigid Inclusion. *Defence Sci. J.* **2007**, *57* (2), 1.
29. Gupta, V. K.; Ray, S.; Singh, S. B. Role of Reinforcement Geometry on the Steady State Creep Behavior of a Rotating Composite Disc. *Multidiscip. Model. Mater. Struct.* **2009**, *5* (2), 139–150.
30. Guven, U. Elastic-Plastic Stresses in a Rotating Annular Disc of Variable Thickness and Variable Density. *Int. J. Mech. Sci.* **1992**, *34* (2), 133–138.
31. Guven, U. The Fully Plastic Rotating Disc of Variable Thickness. *ZAMM* **1994**, *74* (1), 61–65.
32. Guven, U. On the Applicability of Tresca's Yield Condition to the Linear Hardening Rotating Solid Disk of Variable Thickness. *ZAMM* **1995**, *75* (2), 397–398.
33. Heyman, J. Plastic Design of Rotating Discs. *Proc. Inst. Mech. Engrs.* **1958**, *172*, 531–546.

34. Jantinder, K.; Pankaj, T.; Singh, S. B. Steady Thermal Stresses in a Thin Rotating Disc of Finitesimal Deformation with Mechanical Load. *J. Solid Mech.* **2016,** *8* (1), 204–211.

35. Koizumi, M. The Concept of FGM. *Ceram. Trans. Funct. Grad. Mater.* **1993,** *34,* 3–10.

36. Krell, T.; Schulz, U.; Peters, M.; Kaysser, W. A. Graded EB-PVD Alumina Zirconia Thermal Barrier Coatings: An Experimental Approach. *Mater. Sci. Forum* **1999,** 311, 396–401.

37. Låszlö, F. Geschleuderte udrehungskorper im Gebiet bleibender Deformation. *ZAMM* **1925,** *5,* 281–293.

38. Låszlö, F. Rotating Disks in Region of Permanent Deformation. *NACA TM* **1948,** *1192,* 1–27.

39. Lawrence, E. M. *Introduction to the Mechanics of a Continuous Medium;* Prentice-Hall, Inc.: Englewood, NJ, 1962.

40. Lee, W. M. H. Analysis of Plane Stress Problems with Axial Symmetry in Strain Hardening Range. *NACA TM* **1950,** *2217,* 1–79.

41. Leopold, W. R. Centrifugal and Thermal Stresses in Rotating Discs. *Trans. ASME. J. Appl. Mech.* **1948,** *15* (4), 322–326.

42. Ma, B. M. Creep Analysis of Rotating Solid Discs. *J. Franklin Inst.* **1959,** *267,* 149–165.

43. Ma, B. M. A Further Creep Analysis for Rotating Solid Discs of Variable Thickness. *J. Franklin Inst.* **1960,** *269,* 408–419.

44. Ma, B. M. Creep Analysis of Rotating Solid Discs with Variable Thickness and Temperature. *J. Franklin Inst.* **1961,** *271,* 40–55.

45. Ma, B. M. A Power Function Creep Analysis for Rotating Solid Discs Having Variable Thickness and Temperature. *J. Franklin Inst.* **1964,** *277* (6), 593–612.

46. Manson, S. S. The Determination of Elastic Stresses in Gas Turbine Discs. *NACA Rep.* **1947,** *871,* 241–251.

47. Manson, S. S. Direct Method of Design and Stress Analysis of Rotating Discs with Temperature Gradient. *NACA Rep.* **1950,** *952,* 103–116.

48. Manson, S. S. Analysis of Rotating Discs of Arbitrary Contour and Radial Temperature Distribution in the Region of Plastic Deformation. *Proc. U.S. Natl. Congr. Appl. Mech.* **1951,** *1,* 569–577.

49. Millenson, M. B.; Manson, S. S. Determination of Stresses in Gas Turbine Discs Subjected to Plastic Flow and Creep. *NACA Rep.* **1948,** *906,* 277–292.

50. Miyamoto, Y.; Kaysser, W. A.; Rabin, B. H.; Kawasaki, A.; Ford, R. G. *Functionally Graded Materials: Design, Processing, and Applications;* Kluwer Academic Publications: Boston, USA, 1999.

51. Neeraj, C.; Minto, R.; Singh, S. B. Effect of Anisotropy on the Creep of a Rotating Disc of Al–SiCp Composite. *Int. J. Contemp. Math. Sci.* **2010,** *5* (11), 509–516.

52. Neeraj, C.; Minto, R.; Singh, S. B.; Nishi, G. Creep Behavior of Anisotropic Functionally Graded Rotating Discs. *Int. J. Comput. Mater.: Sci. Eng.* **2013,** *12* (1).

53. Odqvist, F. K. G. *Mathematical Theory of Creep and Creep Rupture;* Clarendon Press: Oxford, 1974.

54. Pankaj, T.; Gupta, S. K. Creep Transition in an Isotropic Disc Having Variable Thickness Subjected to Internal Pressure. *Proc. Natl. Acad. Sci. India Sect. A* **2008,** *78* (1), 57.

55. Pankaj, T.; Sonia, R. B. Elastic-Plastic Transition in a Thin Rotating Disc with Inclusion. *World Acad. Sci., Eng. Technol. Int. J. Math. Comput. Sci.* **2008,** *2* (2).

56. Pankaj, T. Elastic-Plastic Transition Stresses in an Isotropic Disc having Variable Thickness Subjected to Internal Pressure. *Int. J. Phys. Sci., Afr. J.* **2009,** *4* (5), 336–342.

57. Pankaj, T. Creep Transition Stresses in a Thin Rotating Disc with Shaft by Finite Deformation under Steady-State Temperature. *Therm. Sci.* **2010,** *14* (2), 425–436.

58. Pankaj, T. Effect of Transition Stresses in a Disc Having Variable Thickness and Poisson's Ratio Subjected to Internal Pressure. *WSEAS Trans. Appl. Theor. Mech.* **2011,** *6* (4), 147–159.

59. Pankaj, T. Elastic-Plastic Transition Stresses in a Thin Rotating Disc with Rigid Inclusion by Infinitesimal Deformation under Steady State Temperature. *Therm. Sci.* **2012,** *14* (1), 209–219.

60. Pankaj, T. Stresses in a Thin Rotating Disc of Variable Thickness with Rigid Shaft. *J. Technol. Plast.* **2012,** *37* (1), 1–14.

61. Pankaj, T.; Singh, S. B.; Jatinder, K. Transitional Stresses Analysis in Thin Rotating Disc for Different Materials. In *Elsevier International Conference on Information and Mathematical Sciences,* 2013; pp 24–26.

62. Pankaj, T.; Singh, S. B.; Jatinder, K. Thickness Variation Parameter in a Thin Rotating Disc by Finite Deformation. *FME Trans.* **2013,** *41* (2), 96–102.

63. Pankaj, T.; Singh, S. B.; Jatinder, K. Steady Thermal Stresses in a Rotating Disc with Shaft Having Density Variation Parameter Subjected to Thermal Load. *Struct. Integrity Life* **2013,** *13* (2), 109–116.

64. Pankaj, T.; Singh, S. B. Elastic-Plastic Transitional Stresses Distribution and Displacement for Transversely Isotropic Circular Disc with Inclusion Subject to Mechanical Load. *Kragujevac J. Sci.* **2015,** *37,* 23–33.

65. Pankaj, T.; Singh, S. B.; Jatinder, K. Mathematical Model in a Thin Non-homogeneous Rotating Disc for Isotropic Material with Rigid Shaft by using Seth's Transition Theory. *Kragujevac J. Sci., Serbia* **2015,** *0.767,* 37.

66. Pankaj, T.; Sandeep, K. Stress Evaluation in a Transversely Isotropic Circular Disk with an Inclusion. *Orig. Sci. Pap. UDK/UDC* **2016,** *539* (37).

67. Pankaja, T.; Gupta, N.; Singh, S. B. Thermal Effect on the Creep in a Rotating Disc by Using Sherby's Law, *Kragujevac J. Sci., Iss.* **2017,** *39,* 17–27.

68. Pankaj, T.; Suresh, K.; Joginder S. Creep Stresses in a Rotating Disc Having Variable Density and Mechanical Load under Steady-State Temperature. *Orig. Sci. Pap. UDK/ UDC: 539.434 Radprimljen,* **2017,** *17* (2), 97–104.

69. Pankaj, T.; Pooja, M.; Sandeep, K. Creep Stresses and Strain Rates for a Transversly Isotropic Disc Having the Variable Thickness under Internal Pressure. *Originalni naucni rad/Orig. Sci. Pap. UDK /UDC: Rad Primljen,* **2017,** *18*(1), 15–21.

70. Pankaj, T.; Shivdev, S.; Nishi G.; Singh, S. B. Effect of Mechanical Load and Thickness Profile on Creep in a Rotating Disc by using Seth's Transition Theory. *J. Am. Inst. Phys.: Conf. Proc. (USA)* **2017,** *1859,* 020024 *[UGC Approved Journal No. 48438].*

71. Pankaj, T.; Sethi, M.; Shivdev, S.; Singh, S. B.; Emmanuel, F. S. Exact Solution of Rotating Disc with Shaft Problem in the Elastoplastic State of Stress having Variable Density and Thickness. *J. Struct. Integr. Life* **2018,** *18* (2), 126–132.

72. Pankaj, T.; Monika, S.; Shivdev, S.; Singh, S. B.; Fadugba, S. E.; Jasmina, L. S. Modeling of Creep Behavior of a Rotating Disc in the Presence of Load and Variable

Thickness by Using Seth Transition Theory. *J. Struct. Integr. Life* **2018**, *18* (2), 133–140 *[UGC Approved Journal No 33851]*.

73. Pankaj, T.; Sandeep, K. Analysis of Stresses in a Transversely Isotropic Thin Rotating Disc with Rigid Inclusion Having Variable Density Parameter. **2013**, *63*, 1–8.

74. Pankaj, T.; Jatinder, K.; Singh, S. B. Thermal Creep Transition Stresses and Strain Rates in a Circular Disc with Shaft Having Variable Density. *Eng. Comput.* **2016**, *33* (3), 698–712.

75. Pankaj, T.; Bansal, S. R. Creep Transition in a Thin Rotating Disc Having Variable Density with Inclusion. *Int. J. Eng. Appl. Sci.* **2007**, *4*, 386–395.

76. Parkus, H. *Thermo-Elasticity*; Springer-Verlag: Wien, New York, 1976.

77. Prandtl, L. Spannungsverteilung in Plastischen Korpern. In *Proceedings of the First. International Congress for Applied Mechanics, Delft – 1924.* Delft 1925, 1924; pp 43–54.

78. Rabotnov, Y. N. Creep Rupture. *Applied Mechanics*; Springer: Berlin-Heidelberg, 1969; pp 342–349.

79. Rabotnov, Y. N. Kinetics of Creep and Creep Rupture. In *Proc. IUTAM Symposia on Irreversibility and Transfer of Physical Characteristics in a Continuum*, Vienna, 1966; pp 325–334.

80. Rees, D. W. A. Elastic-Plastic Stresses in Rotating Discs by von Mises and Tresca, *ZAMM* **1999**, *794*, 281–288.

81. Reuss, A. BerUcksichtigung der elastischen Formanderung in der plastizitatstheorie. *Zeitsch. Angew. Math. Mech.* **1930**, *10*, 266–274.

82. Sahni, M.; Sharma, S. Elastic-Plastic Deformation of a Rotating Solid Disc of Exponentially Varying Thickness and Exponentially Varying Density. In *Proceedings of the International Multi Conference of Engineers and Computer Scientists, Hong Kong II, March 16–18, IMECS* 2016.

83. Sanjeev, S.; Manoj, S. Elastic-Plastic Transition of Transversely Isotropic Thin Rotating Disc. *Contemp. Eng. Sci.* **2009**, *2* (9), 433–440.

84. Sanjeev, S.; Manoj S. Creep Analysis of Thin Rotating Disc Having Variable Thickness and Variable Density with Edge Loading. *Ann. Fac. Eng. Hunedoara—Int. J. Eng. Tome XI Fasc.* **2013**, *3*.

85. Seth, B. R. An Application of the Theory of Finite Strain. *Proc. Math. Sci.* **1939**, *9*, 17–19.

86. Seth, B. R. Transition Theory of Creep Rupture. *Rep. Math. Centre, Univ. Wisconsin*, **1962**, *N.321*, 1–23.

87. Seth, B. R. Transition Theory of Elastic-Plastic Deformation, Creep and Relaxation. *Nature* **1962**, *195*, 896–897.

88. Seth, B. R. The General Measure in Deformation. In *Proc. IUTAM on Second Order Effects in Elasticity, Plasticity and Fluid Dynamics*, Haifa, 1962.

89. Seth, B. R. Asymptotic Phenomenon in Large Rotation. *J. Math. Mech.* **1963**, *12*, 205.

90. Seth, B. R. Generalized Strain and Transition Concepts for Elastic, Plastic Deformation, Creep and Relaxation. *Proc. XIth Int. Congr. Appl. Mech. Munich* 1964; pp 383–389.

91. Seth, B. R. On the Problems of Transition Phenomenon. *Bull. Inst. Politec. Roum.* **1964**, *10*, 255–262.

92. Seth, B. R. Measure Concept in Mechanics. *Int. J. Nonlinear Mech.* **1966**, *I* (2), 35–40.

93. Seth, B. R. Transition Condition: The Yield Condition. *Int. J. Nonlinear Mech.* **1970,** *5,* 279–285.
94. Seth, B. R. Creep Transition. *J. Math. Phys. Sci.* **1972,** *6.*
95. Seth, B. R. Creep Transition in Rotating Cylinder. *J. Math. Phys. Sci.* **1974,** *8* (1), 1–5.
96. Sharma, J. N.; Gupta, S. K.; Pathania, V. *Investigations on the Thermo-Elastic Wave Phenomenon in Anisotropic Material.* PhD Thesis, Department of Mathematics, Himachal Pradesh University, Shimla 17, 2004; pp 100–105.
97. Sharma, S.; Sahni, M. Creep Analysis of Thin Rotating Disc under Plane Stress with No Edge Load. *WSEAS Trans. Appl. Theor. Mech.* **2008,** *7,* 725–738.
98. Sharma, S.; Sahai, I.; Kumar, R. Creep Transition of a Thin Rotating Annular Disc of Exponentially Variable Thickness with Inclusion and Edge Load. *Proc. Eng.* **2013,** *55,* 348–354.
99. Sharma, S.; Sanehlata, Y. Finite Difference Solution of Elastic-Plastic Thin Rotating Annular Disc with Exponentially Variable Thickness and Exponentially Variable Density. *Hindawi Publ. Corp. J. Mater.* Volume 2013, Article ID 809205, 9 pages. http://dx.doi.org/10.1155/2013/809205.
100. Shukla, R. K. Creep Transition in a Thin Rotating Non-homogeneous Disc. *Indian J. Pure Appl. Math.* **1996,** *27* (5), 487–498.
101. Singh, S. B.; Ray, S. Influence of Anisotropy on Creep in a Whisker Reinforced MMC Rotating Disc. In *Proceedings of National Seminar on Composite Materials,* NML Jamshedpur, India, 1998; pp 83–102.
102. Singh, S. B. *Flow Behaviour and Creep Deformation in Engineering Components of Composites.* PhD Thesis, University of Roorkee, 2000.
103. Singh, S. B.; Ray, S. Steady State Creep Behavior in an Isotropic Functionally Graded Material Rotating Disc of Al–SiC Composite. *Metall. Trans. A* **2001,** *32,* 1679–1685.
104. Singh, S. B.; Ray, S. Modeling the Anisotropy and Creep in Orthotropic Aluminum–Silicon Carbide Composite Rotating Disc. *Mech. Mater.* **2002,** *34,* 363–372.
105. Singh, S. B.; Ray, S. Newly Proposed Yield Criterion for Residual Stress and Steady State Creep in an Anisotropic Rotating Composite Rotating Disc. *J. Mater. Proc. Technol.* **2003,** *143–144,* 623–628.
106. Singh, S. B.; Ray, S. Creep Analysis in an Isotropic FGM Rotating Disc of Al–SiC Composite. *J. Mater. Proc. Technol.* **2003,** *143–144,* 616–622.
107. Singh, S. B.; Ray, S. Modeling the Creep in an Isotropic Rotating Disc of Al–SiCw Composite in Presence of Thermal Residual Stress. In *Proc. 3rd International Conference on Advanced Manufacturing Technology: ICMAT-2004,* Kuala Lumpur, Malaysia, 2004; pp 766–770.
108. Singh, S. B. One Parameter Model for Creep in a Whisker Reinforced Anisotropic Rotating Disc of Al–SiCw Composite. *Eur. J. Mech. A: Solids* **2008,** *27,* 680–690.
109. Singh, S. B.; Rattan, M. Creep Analysis of an Isotropic Rotating Al–SiCp Composite Disc Taking into Account the Phase Specific Thermal Residual Stress. *J. Thermoplast. Compos. Mater.* **2010,** *23* (3), 299–312.
110. Sokolniko, I. S. *Mathematical Theory of Elasticity,* 2nd ed.; McGraw-Hill: New York, 1956.
111. Suresh, H. Transition Theory of Creep in Rotating Discs of Variable Thickness. *Indian J. Pure Appl. Math.* **1977,** *9* (3), 597–601.

112. Suresh, H. Thermo-Elastic-Plastic Transition in Rotating Discs with Steady State Temperature, *Indian J. Pure Appl. Math.* **1979**, *10* (5), 597–601.
113. Thompson, A. S. Stresses in Rotating Discs at High Temperature. *Trans. ASME J. Appl. Mech.* **1946**, *13* (1), 45–52.
114. Timoshenko, S. P.; Goodier, J. N. *Theory of Elasticity*, new ed.; McGraw-Hill: New York, 1951.
115. Vandana, G.; Singh, S. B. Creep Analysis in Anisotropic Rotating Disc with Hyperbolically Varying Thickness. In *Proceedings of International Conference on Mechanical and Aerospace Engineering*, SRM University, NCR Campus, Modinagar, 2011; pp 509–513.
116. Vandana, G.; Singh, S. B. Influence of Anisotropy on Creep in an Anisotropic Composite Rotating Disc with Non-Linearly Varying Thickness. *Multidiscip. Model. Mater. Struct.* **2013**, *9*(3), 327–340. https://doi.org/10.1108/MMMS.
117. Vandana, G.; Singh, S. B. Creep Modeling in a Composite Rotating Disc with Thickness Variation in Presence of Residual Stress. *Int. J. Math. Math. Sci.* **2012**, *14*. DOI:10.1155/2012/924921.
118. Vandana, G.; Singh, S. B. Effect of Residual Stress and Reinforcement Geometry in an Anisotropic Composite Rotating Disc Having Varying Thickness. *Int. J. Comput. Mater. Sci. Eng.* **2012**, *01*(04), 1250035. https://doi.org/10.1142/S2047684112500352.
119. Vandana, G.; Singh, S. B. Mathematical Modeling of Creep in a Functionally Graded Rotating Disc with Varying Thickness. *Regen. Eng. Soc. Regen. Eng. Transl. Med.* **2016**, *2*, 126–140.
120. von Mises, R. Mechanics of Solids in the Plastically Deformable State. *NASA, Tech. Memorandom*, **1986**, *1913*, 88488.
121. Wahl, A. M.; Sankey, G. O.; Manjoine, M. J.; Shoemaker, E. Creep Tests of Rotating Discs at Elevated Temperature and Comparison with Theory. *J. Appl. Mech.* **1954**, *21*, 225–235.
122. Wahl, A. M. Analysis of Creep in Rotating Discs Based on the Tresca Criterion and Associated Flow Rule. *J. Appl. Mech.* **1956**, *78*, 231–238.
123. Wahl, A. M. Stress Distributions in Rotating Discs Subjected to Creep at Elevated Temperature. *J. Appl. Mech. ASME Trans.* **1957**, *29*, 299–305.
124. Wahl, A. M. Further Studies of Stress Distribution in Rotating Discs and Cylinders under Elevated Temperature Creep Conditions. *J. Appl. Mech.* **1958**, *80*, 243–250.
125. Wahl, A. M. Effects of Transient Period in Evaluating Rotating Disks Tests under Creep Conditions. *J. Basic Eng.* **1963**, *85*, 66–70.
126. You, L. H.; Long, S. Y.; Zang, J. J. Perturbation Solution of Rotating Solid Discs with Nonlinear Strain Hardening. *Mech. Res. Commun.* **1997**, *24* (6), 649–658.
127. You, L. H.; Zhang, J. J. Elastic-Plastic Stresses in a Rotating Solid Disc. *Int. J. Mech. Sci.* **1999**, *41*, 269–282.
128. You, L. H.; Tang, Y. Y.; Zhang, J. J.; Zheng, C. Y. Numerical Analysis of Elastic-Plastic Rotating Disks with Arbitrary Variable Thickness and Density. *Int. J. Solids Struct.* **2000**, *37*, 7809–7820.

CHAPTER 2

Creep Analysis of Anisotropic Functionally Graded Rotating Disc Subject to Thermal Gradation

MINTO RATTAN[1], TANIA BOSE[1], NEERAJ CHAMOLI,[2] and SATYA BIR SINGH[3*]

[1]Department of Mathematics, UIET, Panjab University, Chandigarh, India

[2]Department of Mathematics, PG DAV College, Chandigarh, India

[3]Department of Mathematics, Punjabi University, Patiala, India

*Corresponding author. E-mail: sbsingh69@yahoo.com

ABSTRACT

The creep behavior of a rotating disc composed of linearly varying functionally graded material subjected to thermal gradation has been studied. This study will provide new insights to material engineers and designers for its applications in gas turbines, jet engines, and other dynamic operators. Rotating disc operates under intense thermomechanical loading, which causes significant creep reducing its service life. The anisotropic yield criterion given by Hill has been used by taking into account different types of anisotropy of disc under thermal gradation. The creep behavior has been described by Sherby's law. Investigations for disc acting under parabolically decreasing temperature profile from inner to outer radii have been carried out. The results obtained for anisotropic discs have also been compared with those obtained for isotropic composites using von Mises criterion. The study revealed that the presence of thermal gradation has reduced the creep rates significantly near the outer radius.

2.1 INTRODUCTION

Aeroengines, compressors, turbines, pumps, flywheels, automobiles, and a number of other dynamic applications have rotating disc as a commonly encountered part.[14] Ma[9] found that the hot gases and conduction from the turbine blades made the blade roots and rim of the disc heated. Hence, these parts are subjected to high temperature and steep thermal gradient than those near the center of the disc. Creep effects are cumulative near the rim of the disc at the same time. Hence, creep of the disc is maximum at the blade roots due to the combined factors of relatively high temperatures, steep thermal gradients and cumulative creep effects. As discs of jet engines and gas turbines are rotated under high temperature and high speeds, the stress produced by the centrifugal forces goes beyond the yield strength of the disc material and hence plastic flow is produced. Thus, composite discs have enormous applications and thermal stress analysis is very important.

Smart materials known as functionally graded materials (FGMs) have continuously graded properties created by nonuniform distributions of reinforcement phase as well as by interchanging the role of reinforcement and matrix materials in a continuous manner. The possibility of tailoring the gradation helps in increasing the performance of FGMs. Metals or ceramics might not sustain alone under high temperature or thermal gradient. Thus, FGMs were invented and helped in the manufacture of high caliber heat-resistant materials.[15] The properties of the material alter due to changes in the composition and structure in FGM. It is used as structural material in high-speed spacecraft and power generation industries as well as many other engineering problems. Metal–ceramic FGMs were used in thermal barrier coatings in aviation and aerospace industry.[11] Singh and Ray[16] carried out secondary creep analysis in a rotating disc made of composites containing SiC_w using Hill yield criterion. It was observed that the tangential stress distribution lowered in the middle of the disc and anisotropy helped to reduce the tangential strain rate significantly, more near inner radius and the strain rate distribution in the orthotropic disc lowered than that in the isotropic disc following von Mises criterion. Further, Singh and Ray[17,18,19] introduced a new yield criterion, which reduced to Hill anisotropic and Hoffman isotropic yield criterion. Rattan et al.[12] studied the secondary creep of thermally graded disc (using isotropic Hoffman yield criterion) rotating at elevated temperature taking into account the phase-specific thermal residual stress. It was observed that there was a significant change in the stress distribution subjected to thermal residual stress. Bose and Rattan[4] studied steady state creep for thermally graded rotating disc having linearly varying FGM and rotating at linearly and parabolically decreasing temperatures. The analysis indicated that stress in FGM disc acting under thermal gradient increased slightly in comparison to disc acting at uniform temperature.

The present chapter formulates a model describing the effect of thermal gradation on the steady state creep in a rotating anisotropic functionally graded disc of aluminum having linearly varying SiC_p from inner to outer radius. Modeling has been done for discs operating at parabolic temperature profile and the results are compared with disc acting at uniform temperature.

2.2   DISC CONFIGURATION

An annular disc of Al (aluminum) matrix reinforced with SiC_p (silicon carbide particles) is considered having inner radius (a) as 31.75 mm; outer radius (b) as 152.4 mm and particle size (p) as 1.7 μ m rotating with angular velocity ω and having uniform thickness h. The dispersed SiC particles vary linearly from the inner to outer radius, so the particle content $vol(r)$ of SiC_p at any radius r is given as:

$$vol(r) = X - Yr, \qquad a \leq r \leq b \tag{2.1}$$

where

$$X = \frac{b.vol_i - a.vol_0}{b - a} \tag{2.2}$$

and

$$X = \frac{vol_i - vol_0}{b - a} \tag{2.3}$$

where vol_i=35.59 vol.% and vol_0=10 vol.% are taken as the particle content at the inner radius and outer radius of the disc.

Law of mixture describes the density variation in the composite disc along the radial distance as

$$\rho(r) = \rho_m + (\rho_d - \rho_m)\frac{vol(r)}{100} \tag{2.4}$$

where densities of Al matrix (ρ_m) and dispersed SiC particles (ρ_d) are 2713 kg/m³ and 3210 kg/m³ respectively.[5] From eqs 2.1 and 2.4, we have

$$\rho(r) = \rho_m + (\rho_d - \rho_m)\frac{X - Yr}{100} \tag{2.5}$$

For vol_{avg} =20 vol.% in the FGM disc, then

$$\int_a^b 2\pi r.h.vol(r)dr = vol_{avg}.\pi.h(b^2 - a^2) \tag{2.6}$$

From eqs 2.1 and 2.6, we get the following relation:

$$vol_{avg} = X - \frac{2Y(b^3 - a^3)}{3(b^2 - a^2)} \tag{2.7}$$

2.3 THERMAL ANALYSIS

A semi-analytical method is employed by dividing the radial domain into finite sub-domains with thickness t_k as done in.[3]

Using curve fitting, temperature distribution in the disc is given by

$$T(r) = 2995r^2 - 1415r + 718.6 \tag{2.8}$$

The below given discs have been considered in the present study;

1. Disc D_1 operating at uniform temperature of 623 K along the radial distance.
2. Disc D_2 operating at parabolically decreasing temperature along the radial distance as per the temperature distribution T(r) mentioned in Ref. [15].

2.4 CREEP PARAMETERS

Sherby's[13] law describes the steady state creep response of the AlSiCp composite as:

$$\dot{\bar{\varepsilon}} = [M(\bar{s} - s_0)]^8 \tag{2.9}$$

where the symbols have their usual meaning.

With the following regression equations, creep parameters M and s_0 as per eq (2.16) are functions of particle size (p), temperature $T(r)$, and volume percent vol(r), and are obtained from the experimental results of Pandey et al.[10]

$$M(r) = p^{0.2112}.T(r)^{4.89}.vol(r)^{-0.591}.e^{-34.91} \qquad (2.10)$$

$$s_0(r) = p^{-0.02050}.T(r)^{0.0378}.vol(r)^{1.033}.e^{-4.9695} \qquad (2.11)$$

2.5   MATHEMATICAL FORMULATION

Principal stresses are in radial, tangential, and axial directions as per symmetry for the AlSiC$_p$ composite disc of uniform thickness, h, rotating with angular velocity, ω Modeling is done with help of the following assumptions:

a) Steady state condition of stress prevails.

b) Elastic deformations being small are neglected in comparison to relatively large creep deformation.

c) At each point of the disc, axial component of stress is assumed to be zero.

d) Small axial movement is allowed as disc is fitted on a splined shaft.

According to Hill's[7] anisotropic yield criterion in principal stress plane and the associated flow rule, the constitutive equations for creep in anisotropic rotating discs take the following form:

$$\dot{\epsilon}_r = \frac{\bar{\dot{\epsilon}}}{2\bar{s}}[(G+H)s_r - Hs_\theta - Gs_z] \qquad (2.12)$$

$$\dot{\epsilon}_\theta = \frac{\bar{\dot{\epsilon}}}{2\bar{s}}[(H+F)s_\theta - Fs_z - Hs_r] \qquad (2.13)$$

$$\dot{\epsilon}_z = \frac{\bar{\dot{\epsilon}}}{2\bar{s}}[(F+G)s_z - Gs_r - Fs_\theta] \qquad (2.14)$$

where $\dot{\epsilon}_r$, $\dot{\epsilon}_\theta$, $\dot{\epsilon}_z$ are the strain rates in the radial, tangential, and axial directions and $\bar{\dot{\epsilon}}$ is the effective strain rate. The effective stress, $\bar{s}$, based on Hill's yield criterion for anisotropic material is given by

$$\bar{s} = \frac{1}{\sqrt{G+H}}[H(s_r - s_\theta)^2 + F(s_\theta - s_r)^2 + G(s_r - s_z)^2] \qquad (2.15)$$

and s_r, s_θ, s_z are stresses in the directions indicated by the subscripts. As per assumption (c), $s_z = 0$, solving constitutive equations along with (2.9) and (2.15) the following relations for strain rates $\dot{\epsilon}_r$, $\dot{\epsilon}_\theta$, and $\dot{\epsilon}_z$ may be obtained as,

$$\dot{\epsilon}_r = \frac{d\dot{u}_r}{dr} = \frac{\sqrt{F(G+H)}\left[\left(\dfrac{G}{F}+\dfrac{H}{F}\right)x-\dfrac{H}{F}\right]}{2\left[\left(\dfrac{G}{F}+\dfrac{H}{F}\right)x^2-2\dfrac{H}{F}x+1+\dfrac{H}{F}\right]^{1/2}}\cdot\left[M(r)(\bar{s}-s_0(r))\right]^8 \quad (2.16)$$

$$\dot{\epsilon}_\theta = \frac{\dot{u}_r}{r} = \frac{\sqrt{F(G+H)}\left[1+\dfrac{H}{F}-\dfrac{H}{F}x\right]}{2\left[\left(\dfrac{G}{F}+\dfrac{H}{F}\right)x^2-2\dfrac{H}{F}x+1+\dfrac{H}{F}\right]^{1/2}}\cdot\left[M(r)(\bar{s}-s_0(r))\right]^8 \quad (2.17)$$

$$\dot{\epsilon}_z = -(\dot{\epsilon}_r+\dot{\epsilon}_\theta) \quad (2.18)$$

where $x = s_r(r)/s_\theta(r)$ at any radius r and u_r is the radial displacement.

The equilibrium equation for the disc of uniform thickness is given by

$$\frac{d}{dr}\left[rs_r(r)\right]-s_\theta(r)+\rho(r)\omega^2 r^2 = 0 \quad (2.19)$$

where the density of the composite is ρ.

Equations 2.16 and 2.17 can be solved to obtain $s_\theta(r)$ as given below:

$$s_\theta(r) = \sqrt{\frac{G+H}{F}}\left[\frac{\sqrt{\dfrac{F}{G+H}}\,s_{\theta_{avg}}(b-a)-\displaystyle\int_a^b \psi_2(r)\,dr}{\displaystyle\int_a^b \psi_1(r)\,dr}\cdot\psi_1(r)+\psi_2(r)\right] \quad (2.20)$$

where

$$\psi_1(r) = \frac{\psi(r)}{2\left[\left(\dfrac{G}{F}+\dfrac{H}{F}\right)x^2-2\dfrac{H}{F}x+1+\dfrac{H}{F}\right]^{1/2}}, \quad (2.21)$$

$$\psi_2(r) = \frac{s_0(r)}{2\left[\left(\dfrac{G}{F}+\dfrac{H}{F}\right)x^2-2\dfrac{H}{F}x+1+\dfrac{H}{F}\right]^{1/2}}, \quad (2.22)$$

$$\psi(r) = \left[\frac{2\left[\left(\dfrac{G}{F}+\dfrac{H}{F}\right)x^2-2\dfrac{H}{F}x+1+\dfrac{H}{F}\right]^{\frac{1}{2}}}{r\sqrt{F(G+H)}\left[1+\dfrac{H}{F}-\dfrac{H}{F}x\right]}\cdot exp\int_a^r \frac{\phi(r)}{r}\,dr\right]^{1/8} \quad (2.23)$$

and

$$\phi(r) = \frac{\left[\left(\dfrac{G}{F} + \dfrac{H}{F}\right)x - \dfrac{H}{F}\right]}{\left[\left(1 + \dfrac{H}{F}\right) - \dfrac{H}{F}x\right]} \tag{2.24}$$

Also, equilibrium eq (2.19) can be solved to obtain the radial stress $s_r(r)$ as

$$s_r(r) = \frac{1}{r}\int_a^r s_\theta(r)\,dr - \frac{\omega^2}{r}\left[\frac{(r^3 - a^3)}{3}\left(\rho_m + (\rho_d - \rho_m)\frac{X}{100}\right) - \frac{(\rho_d - \rho_m)}{100}\frac{Y(r^4 - a^4)}{4}\right] \tag{2.25}$$

After evaluation of $s_\theta(r)$ and $s_r(r)$, strain rates $\dot{\in}_r$, $\dot{\in}_\theta$, and $\dot{\in}_z$ are calculated using eqs (2.12), (2.13), and (2.14) respectively.

An iterative solution technique is employed till the boundary conditions, $s_r(a) = 0$ and $s_r(b) = 0$ are met as in eqs (2.1) and (2.6).

2.6 NUMERICAL CALCULATION

The effect of thermal gradation on the stress and strain rates of anisotropic rotating disc is studied by using the numerical values of anisotropic constants from study of Kulkarni et al.[8] as given in Table 2.1.

TABLE 2.1   Values of Anisotropic Constants.

	Case I	Case II	Case III	Case IV	Case V
$\dfrac{G}{F}$	0.8159	1.2200	1.0000	0.7452	1.3400
$\dfrac{H}{F}$	0.6081	0.7452	1.0000	1.2200	1.6400

The above-mentioned analysis evaluates the stress and strain rate distribution by iterative numerical scheme of computations as given in eq 2.2. The iterations are continued till the process converges. The values of stress and strain rate at different points of the radius grid are obtained.

2.7 VALIDATION OF THE DEVELOPED COMPUTER PROGRAM

The developed computer program is validated and mathematical modeling is done by comparing the results for a rotating steel disc with the available experimental results of eq 2.18 for similar disc and operating under same conditions as done in eq 2.3.

2.7.1 RESULTS AND DISCUSSION

A MATLAB code has been generated to obtain the steady state creep behavior of anisotropic disc subjected to thermal gradation on the basis of mathematical analysis stated earlier.

2.7.2 STRAIN RATES AND STRESS FOR ANISOTROPIC CONSTANTS G/F < 1 AND H/F < 1 (CASE I)

The radial strain rate is compressive throughout the radial distance as it is seen from Figure 2.1(a). At the inner radius (a), the radial strain rate for disc D_2 is $-3.40 \times 10^{-5} s^{-1}$ increases to $-4.89 \times 10^{-5} s^{-1}$ at radius 0.0498 m, decreases to $-3.37 \times 10^{-5} s^{-1}$ at radius 0.0921 m and then finally increases to $-9.19 \times 10^{-5} s^{-1}$ at the outer radius (b) for anisotropic constants (Case I) Disc D_1 attains strain rate of $-1.16 \times 10^{-6} s^{-1}$ at the inner radius and then increases to $-2.28 \times 10^{-3} s^{-1}$. Thus, the radial strain rate is more for disc D_2 near the inner radius and it is least for disc D_2 near the outer radius. The variation of tangential strain rate along radial distance is shown in Figure 2.1(b). The strain rate is $-8.99 \times 10^{-5} s^{-1}$ at the inner radius, increases to $-1.33 \times 10^{-4} s^{-1}$ at radius 0.0438 m, decreases to $-8.92 \times 10^{-5} s^{-1}$ and then gradually increases to $-2.43 \times 10^{-4} s^{-1}$ for disc D_2. However, the strain rate for disc D_1 is $-3.07 \times 10^{-6} s^{-1}$ at the inner radius and then increases to $-6.04 \times 10^{-3} s^{-1}$ at the outer radius. Figure 2.1(c) shows radial stress varies along radial distance in discs D_1 and D_2 operating under the influence of temperature gradient for anisotropic constants (Case I). The graph shows that the maximum value of radial stress for disc D_2 and D_1 are 36.92 MPa and 36.41 MPa respectively at radial distance 0.0800 m. Due to the conditions imposed at the boundary the radial stress is zero at the inner and outer radius. Disc D_1 attains tangential stress of 84.10 MPa at the inner radius, slightly increases to 87.01 MPa at radius 0.0438 m and then

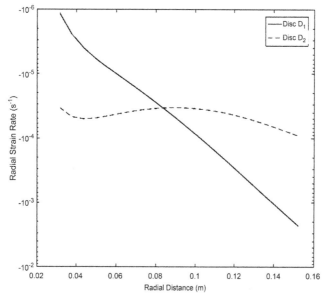

(a)   Radial strain rate versus radial distance.

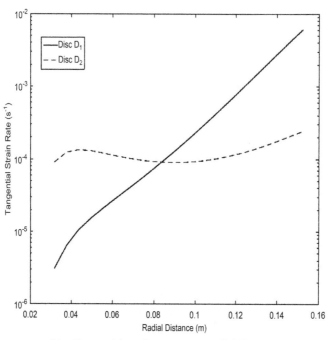

(b)   Tangential strain rate versus radial distance.

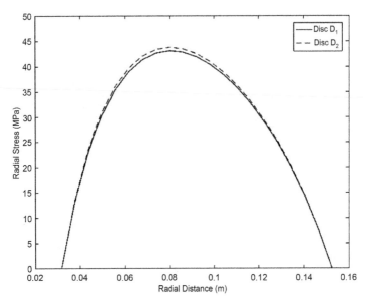

(c)   Radial stress versus radial distance.

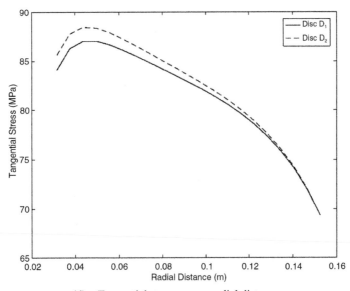

(d)   Tangential stress versus radial distance.

FIGURE 2.1   Representing (a) radial strain rate, (b) tangential strain rate, (c) radial stress, and (d) tangential stress versus radius for discs D_1 and D_2 under the effect of temperature gradient for anisotropic constants (Case I).

decreases to 69.31 MPa at the outer radius for anisotropic constants (Case I). However, disc D_2 attain values 85.62 MPa at the inner radius, increases to 88.42 MPa at radial distance 0.0438 m and then decreases to the same value as disc D_1 as seen in Figure 2.1(d).

2.7.3   STRAIN RATES AND STRESS FOR ANISOTROPIC G/F=1 AND H/F=1 (CASE III)

The radial strain rate is compressive throughout the radial distance as it could be seen from Figure 2.2(a). At the inner radius, the radial strain rate for disc D_2 is -2.72×10^{-5} s^{-1}, increases to -7.04×10^{-3} s^{-1} at radius 0.0559 m, decreases to -5.76×10^{-5} s^{-1} at radius 0.0921 m and then finally increases to -8.85×10^{-5} s^{-1} at the outer radius for anisotropic constant (Case III) Disc D_1 attains strain rate -9.17×10^{-7} s^{-1} at the inner radius and increases till it attains strain rate of -2.18×10^{-3} s^{-1} at the outer radius. Thus, the radial strain rate is more for disc D_2 near the inner radius and it is least for the same disc near the outer radius. The variation of tangential strain rate along radial distance for anisotropic constant (Case III) is shown in Figure 2.2(b). The strain rate for disc D_2 is -1.83×10^{-3} s^{-1} at the inner radius and gradually increases to -4.36×10^{-3} s^{-1} at the outer radius. The maximum value of radial stress for disc D_1 and D_2 are 39.48 MPa and 40.07 MPa respectively at radial distance 0.0800 m as seen in Figure 2.2(c) for anisotropic constant (Case III). The radial stress follows a parabolic path as it attains zero stress value at the inner and outer radius due to imposed boundary conditions. Figure 2.2(d) displays the variation of tangential stress along radial distance in discs D_1 and D_2 operating under the influence of temperature gradient for anisotropic constant (Case III). The discs D_1 and D_2 attain maximum tangential stress of 91.13 MPa and 92.51 MPa at radial distance 0.0559 m and then decreases to 69.31 MPa at the outer radius.

2.7.4   STRAIN RATES AND STRESS FOR ANISOTROPIC G/F>1 AND H/F>1 (CASE V)

Figure 2.3(a) reveals that the radial strain rate is compressive throughout the radial distance. At the inner radius, the radial strain rate for disc D_2 is -2.09×10^{-5} s^{-1}, increases to -1.17×10^{-3} s^{-1} at radius 0.0619 m, decreases

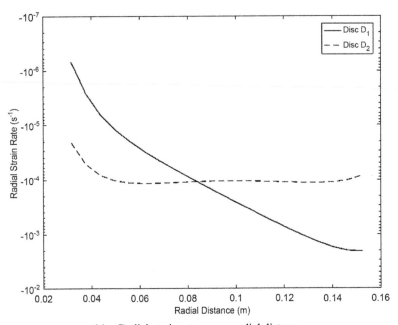

(a)   Radial strain rate versus radial distance.

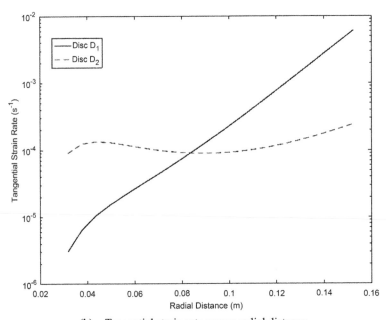

(b)   Tangential strain rate versus radial distance.

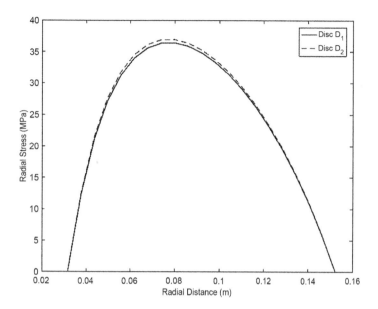

(c)   Radial stress versus radial distance.

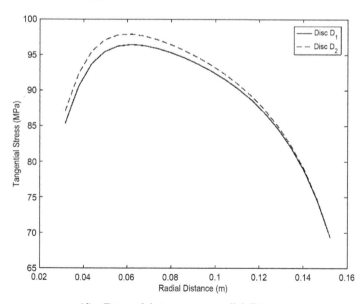

(d)   Tangential stress versus radial distance.

FIGURE 2.2   Representing (a) radial strain rate, (b) tangential strain rate, (c) radial stress, and (d) tangential stress versus radius for discs D_1 and D_2 under the effect of temperature gradient for anisotropic constants (Case III).

to -1.05×10^{-4} s^{-1} at radius 0.0981 m, again increases to -1.14×10^{-4} s^{-1} at radius 0.1343 m, and finally decreases to -8.43×10^{-5} s^{-1} at the outer radius for anisotropic constants (Case V). Thus, graph shows the radial strain rate is more for disc D_2 near the inner radius and it is least for the same disc near the outer radius. The variation of tangential strain rate along radial distance is shown in Figure 2.3(b). The strain rate is 3.37×10^{-5} s^{-1} at the inner radius, increases to 1.88×10^{-4} s^{-1} at radius 0.0619 m, decreases to 1.70×10^{-4} s^{-1} at radius 0.0981 m, again increases to 1.84×10^{-4} s^{-1} at radius 0.1343 m, and finally decreases to 1.35×10^{-4} s^{-1} at the outer radius. However, the strain rate for disc D_1 is 1.11×10^{-3} s^{-1} at the inner radius and then gradually increases to 3.32×10^{-3} s^{-1} at the outer radius. Figure 2.3(c) shows radial stress varies along radial distance in discs D_1 and D_2 operating under the influence of temperature gradient for anisotropic constants (Case V). From the graph it is depicted that the maximum value of radial stress for disc D_2 and D_1 are 43.77 MPa and 43.09 MPa respectively at radial distance 0.0800 m. The discs D_1 and D_2 attain tangential stress 85.33 MPa and 87.07 MPa at the inner radius, increases to 96.40 MPa and 97.85 MPa, respectively at radial distance 0.0619 m and then merges to 69.31 MPa at the outer radius as shown in Figure 2.3(d).

2.8   CONCLUSIONS

The following conclusions are drawn in the context of the present study. The radial stress increases from zero at the inner radius, reaches maximum before vanishing to zero at the outer radius under the boundary conditions imposed at the inner and outer radii. The radial stress in the rotating disc increases throughout the radial distance under the effect of uniform and parabolic temperature distribution for all the anisotropic constants discussed. The incorporation of parabolic temperature profile on the FGM disc led to increase in the tangential stress near the inner radius but as one move toward the outer radius, the difference in the stress values for both the discs goes on decreasing and finally merges at the outer radius. In FGM discs, the strain rates tend to become relatively more uniform over the entire radius of the disc with the imposition of linear and parabolic temperature distribution. Consequently, the study of thermal gradation in anisotropic functionally graded rotating discs has been made to give new insights to material engineers and designers for applications in gas turbines, jet engines and other dynamic operators.

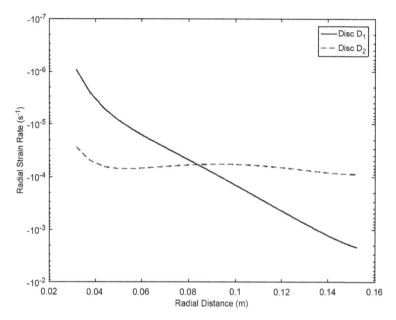

(a)   Radial strain rate versus radial distance.

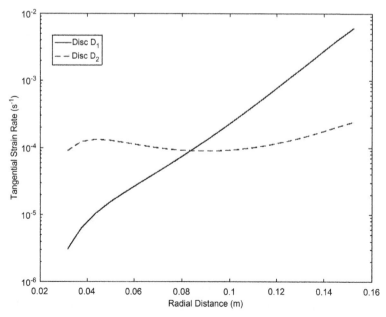

(b)   Tangential strain rate versus radial distance.

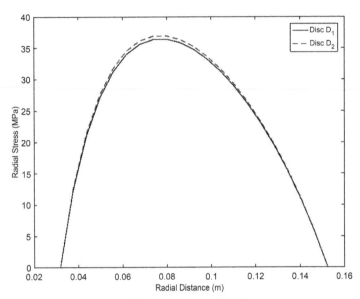

(c)   Radial stress versus radial distance.

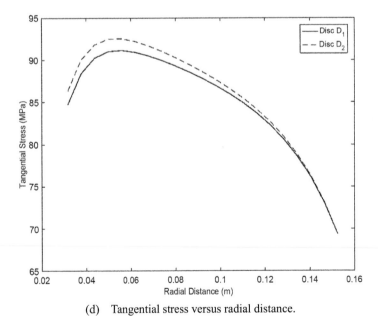

(d)   Tangential stress versus radial distance.

FIGURE 2.3   Representing (a) radial strain rate, (b) tangential strain rate, (c) radial stress, and (d) tangential stress versus radius for discs D_1 and D_2 under the effect of temperature gradient for anisotropic constants (Case V).

ACKNOWLEDGMENTS

The authors whole heartedly thank Council of Scientific and Industrial Research, Human Resource Development Group, New Delhi, India for the grant (F. No. 09/135(0706)/2014-EMR-I).

KEYWORDS

- **creep**
- **anisotropy**
- **rotating disc**
- **thermal gradation**

REFERENCES

1. Arnold, S. M.; Saleeb, A. F.; Al-Zoubi, N. R. Deformation and Life Analysis of Composite Flywheel Disk and Multi-Disk Systems. NASA/TM-2001-210578: 2001; 1–50.
2. Bose, T.; Rattan, M. Modeling Creep Analysis of Thermally Graded Anisotropic Rotating Composite Disc. *Int. J. App. Mech.* **2018,** *10* (6): 1850063(15 pages).
3. Bose, T.; Rattan, M. Effect of Thermal Gradation on Steady State Creep of Functionally Graded Rotating Disc. *Eur. J. Mech. A/Solids* **2018,** *67*; 169–176.
4. Bose, T.; Rattan, M. Modeling Creep Behavior of Thermally Graded Rotating Disc of Functionally Graded Material. Differential Equations and Dynamical Systems, DOI: 10.1007/s12591-017-0350-1. 2016.
5. Clyne, T. W.; Withers, P. J. An Introduction to Metal Matrix Composites; Cambridge University Press: Cambridge, 1993; p. 479.
6. Garg, M.; Salaria, B. S.; Gupta, V. K. Effect of Thermal Gradient on Steady State Creep in a Rotating Disc of Variable Thickness. *Procedia Eng.***2013,** *55*; 542–547.
7. Hill, R. A Theory of the Yielding and Plastic Flow of Anisotropic Metals. *P. Roy. Soc. A- Math. Phy..* **1948,** *193* (1033); pp. 281–297.
8. Kulkarni, P. S.; Bhatnagar, N. S.; Arya, V. K. Creep Analysis of Thin-Walled Anisotropic Cylinders Subjected to Internal Pressure, Bending and Twisting. Proceedings of the Workshop on Solid Mechanics, University of Roorkee, 1985.
9. Ma, B. M. Creep Analysis of Rotating Disks with Variable Thickness and Temperature. *J. Franklin Inst.* **1961,** *271*, 40–55.

10. Pandey, A. B.; Mishra, R. S.; Mahajan, Y. R. Steady State Creep Behavior of Silicon Carbide Particulate Reinforced Aluminum Composites. *Acta Metall. Mater.* **1992,** *40* (8), 2045–2052.

11. Rattan, M.; Chamoli, N.; Singh, S. B.; Gupta, N. Creep Behavior of Anisotropic Functionally Graded Rotating Discs. *Int. J. Comput. Mat.* **2013,** *2* (1), 1350005–1–15.

12. Rattan, M.; Kaushik A.; Chamoli, N.; Bose, T. Steady State Creep Behavior of Thermally Graded Isotropic Rotating Disc of Composite Taking into Account the Thermal Residual Stress. *Eur. J. Mech. A/Solids* **2016,** *60,* 315–326.

13. Sherby, O. D.; Klundt, R. H.; Miller, A. K. Flow Stress, Subgrain Size and Subgrain Stability at Elevated Temperature. *Metall. Trans. A.* **1977,** *8,* 843–850.

14. Singh, S. B. One Parameter Model for Creep in a Whisker Reinforced Anisotropic Rotating Disc of Al-SiCw Composite. *Eur. J. Mech. A/Solids* **2008,** *27,* 680–690.

15. Singh, T.; Gupta, V. K. Analysis of Steady State Creep in Whisker Reinforced Functionally Graded Thick Cylinder Subjected to Internal Pressure by Considering Residual Stress. *Mech. Adv. Mat. Struc.* **2014,** *21,* 384–392.

16. Singh, S. B.; Ray, S. Newly Proposed Yield Criterion for Residual Stress and Steady State Creep in An Anisotropic Composite Rotating Disc. *J. Mater. Process. Technol.* **2003,** *143–144,* 623–628.

17. Singh, S. B.; Ray, S. Modeling the Anisotropy and Creep in Orthotropic Aluminum–Silicon Carbide Composite Rotating Disc. *Mech. Mater.* **2002,** *34* (6), 363–372.

18. Singh, S. B.; Ray, S. Steady State Creep Behavior in An Isotropic Functionally Graded Material Rotating Disc of Al-SiC Composite. *Metall. Mater. Trans. A* **2001,** *32* (7), 1679–1685.

19. Singh, S. B.; Ray, S.; Gupta, R. K.; Bhatnagar, N. S. Influence of Anisotropy on Creep in a Whisker Reinforced MMC Rotating Disc, Proceedings composite materials COMPEAT-1998, National Metallurgical Laboratory, Material Research Society of India, 1998; pp. 83–102.

20. Wahl, A. M.; Sankey, G. O.; Manjoine, M. J.; Shoemaker, E. Creep Tests of Rotating Disks at Elevated Temperature and Comparison with Theory. *J. App. Mech.* **1954,** *76,* 225–235.

CHAPTER 3

Elastic-Plastic Transitional Stress Analysis of Human Tooth Enamel and Dentine Under External Pressure Using Seth's Transition Theory

SHIVDEV SHAHI[1*] and S. B. SINGH[2*]

1,2Department of Mathematics, Punjabi University Patiala, Punjab, India

Corresponding author. E-mail: shivdevshahi93@gmail.com; sbsingh69@yahoo.com

ABSTRACT

In this paper, elastic–plastic transitional stresses in human tooth enamel and dentine are calculated analytically. The tooth is modelled in the form of a shell, which exhibits transversely isotropic macrostructural symmetry. Transition theory of B. R. Seth has been used to model the elastic–plastic state of stresses. The shell so modeled is subjected to external pressure to analyze the state of stress. The results for enamel and dentine are compared with hydroxyapatite (HAP), given by chemical formula $Ca_{10}(PO_4)_6(OH)_2$. It is a naturally occurring mineral form of calcium and the enamel is 95% wt. composed of this mineral. The elastic stiffness constants for these are taken from the available literatures that have been obtained using resonance spectroscopy, a nondestructive technique of obtaining the elasticity constants. The radial and circumferential stresses are obtained for radius ratios, which can handle any type of dataset for thicknesses of enamel and dentine.

3.1 INTRODUCTION

Stresses that must sustain the human mastication system originate in the intermaxilar contact points and propagate through the dentine and enamel.[16] Dental enamel is the hardest tissue in the human body, which protects reliably a tooth from mechanical loading and aberration. It is a highly mineralized substance that contains ~96% apatite by weight.[3] One of the most important crystal that comprise dental enamel is HAP, given by chemical formula $Ca_{10}(PO_4)_6(OH)_2$. According to early works, it has been concluded that dental enamel exhibits low ability to deformation prior the failure at ultimate compressive strength ~100–400 MPa.[8,9,15] Human dentine is the internal structural bulk of the tooth below the enamel. Dentine is likewise a biological compound formed by ~50% carbonated apatite mineral. As a stereotype, dentine is also characterized as a brittle substance, but it was recently shown that human dentine can demonstrate high elastic and plastic limits at ultimate compressive strengths ~800 MPa.[4,7,10] This paper also concerns with the calculation of elastic–plastic transition stress concentrations in enamel and dentine using Seth's transition theory. The tooth will be modeled as a spherical shell exhibiting transversely isotropic macrostructural symmetry. Equations for modeling spherical shells made of isotropic materials are available in most standard text books.[2,6,11,12,14,24] Miller evaluated solutions for stresses and displacements in a thick spherical shell subjected to internal and external pressure loads.[13] You et al. (2005) presented a highly precise model to carry out elastic analysis of thick-walled spherical pressure vessels.[25] The authors have studied the behavior of shells particularly when some assumptions, such as (1) incompressibility of material used, (2) creep strain law derived by Norton, (3) yield condition of Tresca, and (4) associated flow rules were made. The need of utilization of these specially appointed semi-experimental laws in elastic–plastic transition depends on the approach that the transition is a linear phenomenon that is unrealistic. Deformation fields related with irreversible phenomenon, such as elastic–plastic disfigurements, creep relaxation, fatigue and crack, and so forth are nonlinear in character. The traditional measures of deformation are not adequate to manage transitions. The concept of generalized strain measures and transition theory given by Seth (1962) has been applied to find elastic–plastic stresses in various problems by solving the nonlinear differential equations at the transition points.

Thakur successfully analyzed creep transition stresses of a thick isotropic spherical shell by finitesimal deformation under steady state of temperature and internal pressure by using Seth's transition theory.[22] The theory has been used to solve various problems of stress and strain determination in structures modeled in the form of disc.[20,23] All these problems were based on the recognition of the transition state as separate state necessitates showing the existence of the used constitutive equation for that state.

3.2 GOVERNING EQUATIONS

We consider a spherical shell of constant thickness with internal and external radius a and b respectively under external pressure p. The geometry of the shell is symmetric in nature as shown in Figure 3.1.

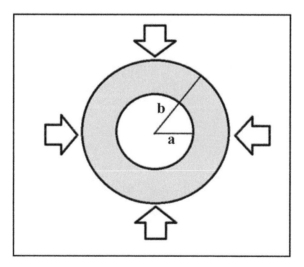

FIGURE 3.1   Geometry of spherical shell under the influence of external pressure p.

Displacement coordinates: The components of displacement in spherical coordinates (r,θ,ϕ) are taken as:

$$u = r(1-\beta); v = 0; w = 0 \tag{3.1}$$

where β is position function depending on r.

The generalized components of strain are given by B. R. Seth (1962, 1966) as:

$$e_{rr} = \frac{1}{n}\left[1-(r\beta'+\beta)^n\right] = \frac{1}{n}\left[1-\beta^n(1+P)^n\right]$$

$$e_{\theta\theta} = e_{\varphi\varphi} = \frac{1}{n}\left[1-\beta^n\right],$$

$$e_{r\theta} = e_{\theta\varphi} = e_{\varphi r} = 0. \qquad (3.2)$$

where n is the measure and $r\,\beta' = \beta P$

P is a function of β and β is a function of r.

Stress–strain relation: The stress–strain relations for isotropic material are given by Sokolnikoff (1956):

$$T_{ij} = c_{ijkl}\,e_{kl},\ (i,j,k,l = 1,2,3)$$

where T_{ij} and e_{kl} are the stress and strain tensors respectively. These nine equations contain a total of 81 coefficients e_{ijkl}, but not all the coefficients are independent. The symmetry of T_{ij} and e_{ij} reduces the number of independent coefficients to 36. For transversely isotropic materials that have a plane of elastic symmetry, these independent coefficients reduce to 5. The constitutive equations for transversely isotropic media are given by Altenbach et al. (2004):

$$
\begin{bmatrix} T_{11} \\ T_{22} \\ T_{33} \\ T_{23} \\ T_{31} \\ T_{12} \end{bmatrix} =
\begin{bmatrix}
c_{11} & c_{12} & c_{13} & 0 & 0 & 0 \\
c_{12} & c_{11} & c_{13} & 0 & 0 & 0 \\
c_{13} & c_{13} & c_{33} & 0 & 0 & 0 \\
0 & 0 & 0 & c_{44} & 0 & 0 \\
0 & 0 & 0 & 0 & c_{44} & 0 \\
0 & 0 & 0 & 0 & 0 & \frac{1}{2}(c_{11}-c_{12})
\end{bmatrix}
\begin{bmatrix} e_{11} \\ e_{22} \\ e_{33} \\ e_{23} \\ e_{31} \\ e_{12} \end{bmatrix}, \qquad (3.3)
$$

Substituting eq (3.2) in eq (3.3), we get

$$T_{rr} = \frac{c_{33}}{n}\left[1-(r\beta'+\beta)^n\right]+\frac{1}{n}(2c_{12})(1-\beta^n) \Rightarrow c_{33}e_{rr}+2c_{12}e_{\theta\theta};$$

$$T_{\theta\theta} = T_{\varphi\varphi} = \frac{c_{21}}{n}\left[1-(r\beta'+\beta)^n\right]+\frac{2}{n}(c_{11}-c_{66})(1-\beta^n) \Rightarrow c_{12}e_{rr}+2(c_{11}-c_{66})e_{\theta\theta};$$

$$T_{r\theta} = T_{\theta\phi} = T_{\phi r} = 0 \qquad (3.4)$$

Equation of equilibrium: The equations of equilibrium are:

$$\frac{\partial T_{rr}}{\partial r} + \frac{1}{r\sin\theta}\frac{\partial T_{r\varphi}}{\partial\varphi} + \frac{1}{r}\frac{\partial T_{r\theta}}{\partial\theta} + \frac{2T_{rr} - T_{\theta\theta} - T_{\varphi\varphi} + T_{r\theta}\cot\theta}{r} = 0;$$

$$\frac{\partial T_{r\theta}}{\partial r} + \frac{1}{r\sin\theta}\frac{\partial T_{\theta\phi}}{\partial\varphi} + \frac{1}{r}\frac{\partial T_{\theta\theta}}{\partial\theta} + \frac{3T_{r\theta} + \left(T_{\theta\theta} - T_{\phi\phi}\right)\cot\theta}{r} = 0;$$

$$\frac{\partial T_{r\phi}}{\partial r} + \frac{1}{r\sin\theta}\frac{\partial T_{\phi\phi}}{\partial\phi} + \frac{1}{r}\frac{\partial T_{\phi\theta}}{\partial\theta} + \frac{3T_{r\phi} + 2T_{\theta\theta}\cot\theta}{r} = 0. \tag{3.5}$$

Substituting eq (3.4) in eq (3.5), we see that the equations of equilibrium are all satisfied except:

$$\frac{\partial T_{rr}}{\partial r} + \frac{2T_{rr} - T_{\theta\theta} - T_{\phi\phi}}{r} = 0 \tag{3.6}$$

$$or\quad \frac{\partial T_{rr}}{\partial r} + \frac{2}{r}(T_{rr} - T_{\theta\theta}) = 0 \tag{3.7}$$

From eq (3.7), one may also say that

$$T_{\phi\phi} - T_{\theta\theta} = 0 \tag{3.8}$$

Equation (3.8) is satisfied by $T_{\theta\theta}$ and $T_{\phi\phi}$ as given by eq (3.2). If $c_{21} = c_{31}$, $c_{22} - c_{33} = c_{32} - c_{23}$ the equation of equilibrium from eq (3.6) becomes:

$$\frac{\partial T_{rr}}{\partial r} + \frac{2\left(T_{rr} - T_{\theta\theta}\right)}{r} = 0, \tag{3.9}$$

Critical points or turning points: By substituting eq (3.4) into eq (3.9), we gets a nonlinear differential equation in terms of β:

$$P(P+1)^{n+1}\beta\frac{dP}{d\beta} + P(P+1)^n + 2(1-C_1)P - \frac{2}{n\beta^n}[Q_1 - Q_2] \tag{3.10}$$

where $Q_1 = C_1\{1 - \beta^n(1-P)^n\}$, $Q_2 = C_2(1-C_1)\{1-\beta^n\}$, $C_1 = (c_{33} - c_{13})/c_{33}$ and $C_2 = (c_{11} + c_{12} - 2c_{13})/(c_{33}(1-C_1))$

P is function of β and β is function of r only.

Transition points: The transition points of β in eq (3.10) are $P = 0$, $P \to -1$ and $P \to \pm\infty$.

To solve the elastic–plastic stress problems we consider the case of $P \to \pm\infty$.

Boundary condition: The boundary conditions of the problem are given by

$$r = a, \tau_{rr} = 0$$
$$r = b, \tau_{rr} = -p \tag{3.11}$$

3.3 PROBLEM SOLUTION

For finding the elastic–plastic stresses, the transition function is taken through the principal stresses at the transition point $P \to \pm \infty$, we define the transition function ζ as:

$$\zeta = (3 - 2C_1) - \frac{nT_{rr}}{(c_{33})} \cong \left[\beta^n (P+1)^n + 2(1 - C_1) \right] \tag{3.12}$$

where ζ be the transition function unction of r only. Taking the logarithmic differentiation of eq (3.12), with respect to r and using eq (3.10), we get

$$\frac{d(\log \zeta)}{dr} = \frac{-2C_1}{r} \tag{3.13}$$

Taking the asymptotic value of eq (3.13) as $P \to \pm \infty$ and integrating, we get

$$\zeta = Ar^{-2C_1} \tag{3.14}$$

where A is a constant of integration and $C_1 = (c_{33} - c_{13}) / c_{33}$. From eq (3.12) and (3.14), we have

$$T_{rr} = \frac{c_{33}}{n} \left[(3 - 2C_1) + Ar^{-2C_1} \right] \tag{3.15}$$

Using boundary condition from eq (3.11) in eq (3.15), we get

$$A = -\frac{(3 - 2C_1)}{a^{-2C_1}} \quad \text{and} \quad p = -\frac{c_{33}}{n} \left[(3 - 2C_1) \left[1 - \left(\frac{b}{a} \right)^{-2C_1} \right] \right] \tag{3.16}$$

Substituting eq (3.16) into eq (3.15) and using eq (3.16) in equation of equilibrium, we get

$$T_{rr} = \frac{c_{33}}{n} \left[(3 - 2C_1) \left[1 - \left(\frac{r}{a} \right)^{-2C_1} \right] \right]$$

$$T_{\theta\theta} = T_{rr} + \frac{c_{33}}{n} \left[(3 - 2C_1)C_1 \left(\frac{r}{a} \right)^{-2C_1} \right] \tag{3.17}$$

Initial Yielding: From eq (3.17), it is seen that $|T_{\theta\theta} - T_{rr}|$ is maximum at the outer surface (that is at $r = b$), therefore yielding of the shell will take place at the external surface of the shell:

$$|T_{\theta\theta} - T_{rr}|_{r=b(external\ surface)} = \left| \frac{c_{33}}{n} \left[(3 - 2C_1)C_1 \left(\frac{b}{a} \right)^{-2C_1} \right] \right| \equiv Y\ (Yielding\ )\quad (3.18)$$

Using eq (3.18) in eqs (3.15) to (3.17), we get the transitional stresses as in nondimensional components as:

$$\sigma_{rr} = \frac{1}{C_1} \frac{\left(1 - R^{-2C_1} \right)}{\left(R_0^{-2C_1} \right)}; \sigma_{\theta\theta} = \frac{1}{C_1} \frac{\left(1 - R^{-2C_1} \right)}{\left(R_0^{-2C_1} \right)} + \left(\frac{R^{-2C_1}}{R_0^{-2C_1}} \right)\ \text{and}\ P_{oi} = \frac{\left(R_0^{-2C_1} - 1 \right)}{C_1 R_0^{-2C_1}}$$

where $R = r/a$, $R_0 = b/a$, $\sigma_{rr} = T_{rr}/Y$, $\sigma_{\theta\theta} = T_{\theta\theta}/Y$ and $P_{oi} = p/Y$ (3.19)

Fully Plastic state: For fully plastic case [Seth, 1963], $C_1 \to 0$; therefore, stresses and pressure from eq (3.19) becomes

$$\sigma_{rr} = 2Y^* \log R_0; \sigma_{\theta\theta} = 2Y^* \log R\ \text{and}\ P_{of} = Y^* + \sigma_{rr} \quad (3.20)$$

where $R = r/a$, $R_0 = b/a$, $\sigma_{rr} = T_{rr}/Y^*$, $\sigma_{\theta\theta} = T_{\theta\theta}/Y^*$ and $P_{of} = p/Y^*$

3.4   NUMERICAL RESULTS AND DISCUSSION

The above-mentioned investigations elaborate the initial yielding and fully plastic state of enamel and dentine modeled in form of spherical shell subjected to external pressure. The elastic constants for enamel and dentine are taken from the literature,[5] which have been obtained by ultrasonic resonance spectroscopy, a nondestructive measure to obtain the stiffness constants. The results of enamel and dentine are calculated along with HAP, which is an integral part of enamel and dentine and is also used in the manufacture of dental implants. All these materials exhibit transversely isotropic macrostructural symmetry. In Figure 3.2 the curves are plotted for pressure at initial yielding at various radius ratios. It is observed that intensity of pressure at initial yielding increases with increase in thickness of the shell. Dentine being the softer tissue, yielded at lower levels of stress as compared to enamel and HAP. Figures 3.3 and 3.4 show the trends of radial and circumferential stresses at initial yielding. Maximum stresses were observed at external surface of the shell. In Figure 3.5 the curves are plotted for stresses required at fully plastic state for various radius ratios. It has been observed that shells exhibited high plasticity when the thickness of the shell was between ratios $1 < R_0 < 2$. Significant drop in the levels of plasticity is observed when the thickness increases. The plasticity of dentine is inferred to greater than that of enamel. Figures 3.6 and 3.7 represent the trends of radial and circumferential stresses at fully plastic state. The observations infer to the fact that the principal stress differences were maximum at external surface of the shells. Stresses at fully plastic state for HAP and enamel were greater than stresses in dentine. HAP being the stiffest of the three, required highest levels of stress for yielding at various radius ratio.

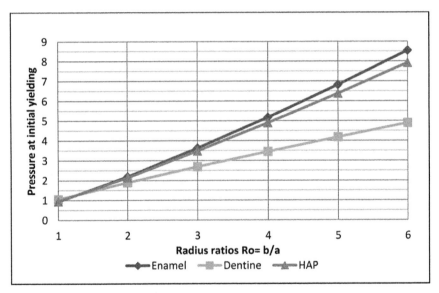

FIGURE 3.2   Pressure in the Shell at initial yielding for radius ratio R_0.

3.5   CONCLUSIONS

On the basis of numerical results, it can be concluded that enamel and dentine under uniaxial compression behave like a functionally graded strong hard tissue with necessary elastic and plastic limit. It demonstrates considerable ability to suppress a crack growth. Different values of

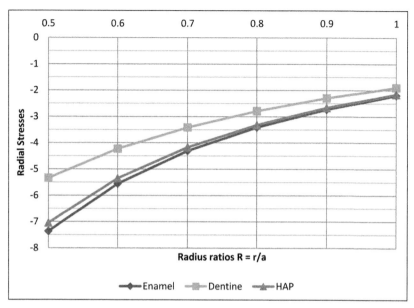

FIGURE 3.3 Radial stresses at initial yielding.

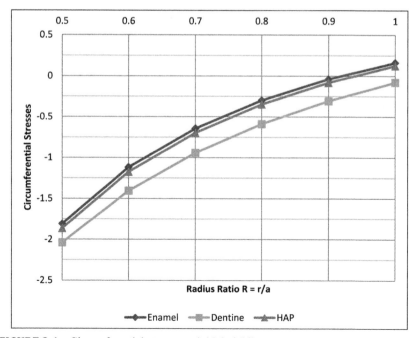

FIGURE 3.4 Circumferential stresses at initial yielding.

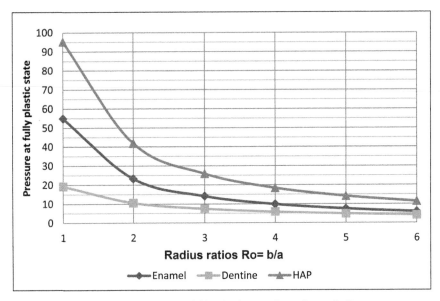

FIGURE 3.5   Pressure in the shell at fully plastic state for radius ratio R_0.

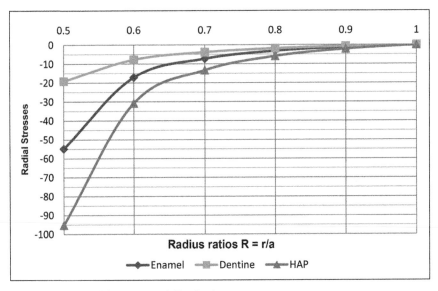

FIGURE 3.6   Radial stresses at fully plastic state.

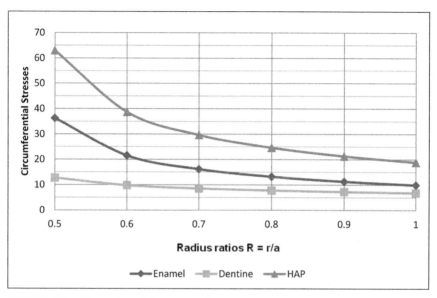

FIGURE 3.7   Circumferential stresses at fully plastic state.

pressure required for initial yielding and fully plastic state were calculated for various radius ratios depending on the geometry of the sample. Trends of the graphs were similar for enamel and hydroxyapatite due to enamel's composition. Significant difference between stress buildup at crown and root dentine has been observed by varying the radii ratios in the modeled spherical shell. The mathematical model developed to analyze the tranversely isotropic behavior of enamel and dentine may be of great use in the manufacture of highly competitive dental implants.

KEYWORDS

- **enamel**
- **dentine hydroxyapatite**
- **shell**
- **stresses**
- **transverse isotropy**

REFERENCES

1. Altenbach, H.; Altenbach, J.; Kissing, W. *Mechanics of Composite Structural Elements*; Springer-Verlag: Berlin Heidelberg, 2004.
2. Boyle, J. T.; Spence, J. *Stress Analysis for Creep*; Butterworths Coy. Ltd: London, 1983.
3. Cuy, J. L.; Mann, A. B.; Livi, K. J.; Teaford, M. F.; Weihs, T. P.; Nanoindentation Mapping of the Mechanical Properties of Human Molar Tooth Enamel. *Arch. Oral Biol* 2002, *47*, 281–291.
4. Zaytsev, D.; Grigoriev, S.; Panfilov, P. Deformation Behavior of Human Dentine under Uniaxial Compression. *Int. J. Biomater.* 2012, (*2012*, Article ID 854539, 8 pages).
5. Menéndez-Proupina, E.; Cervantes-Rodríguezb, S.; Osorio-Pulgara, R.; Franco-Cisternaa, M.; Camacho-Montesc, H.; Fuentesb, M. E. Computer Simulation of Elastic Constants of Hydroxyapatite and Fluorapatite. *J. Mech. Behav. Biomed. Mater.* 2011, *4*, 1011–1020.
6. Fung, Y. C. *Foundations of Solid Mechanics*; Englewood Cliffs, N. J.: Prentice-Hall, 1965.
7. Kinney, J. H.; Balooch, M.; Marshall, S. J.; Marshall, G. W. Amicromechanics Model of the Elastic Properties of Human Dentine. *Arch. Oral Biol.* 1999, *44*, 813–822.
8. Stanford, J. W.; Weigel, K. V.; Paffenberger, G. C.; Sweeney, W. T. Compressive Properties of Hard Tooth Tissues and Some Restorative Materials. *J. Am. Dent. Assoc.* 1960, *60*, 746–751.
9. Stanford, J. W.; Weigel, K. V.; Paffenberger, G. C.; Sweeney, W. T. Determination of Some Compressive Properties of Human Enamel and Dentine. *J. Am. Dent. Assoc.* 1958, *57*, 487–495.
10. Liang, J. Z. Toughening and Reinforcing in Rigid Inorganic Particulate Filled Poly (propylene): A Review. *J. Appl. Polym. Sci.* 2001, *83* (7), 1547–1555.
11. Kraus, H. *Creep Analysis*; Wiley; New York, USA, 1980.
12. Lubhan, D.; Felger, R. P. *Plasticity and Creep of Metals*; Wiley: New York, USA, 1961.
13. Miller, G. K. Stresses in a Spherical Pressure Vessel Undergoing Creep and Dimensional Changes. *Int. J. Sol. Struct.* 1995, *32*, 2077–2093.
14. Parkus, H. *Thermo - Elasticity*; Springer-Verlag: Wien, New York, 1976.
15. Craig, R. G.; Peyton, F. A.; Johnson, D. W. Compressive Properties of Enamel, Dental Cements, and Gold. *J. Dent. Res.* 1961, *40*, 936–945.
16. Marshall, S. J.; Balooch, M.; Breunig, T.; Kinney, J. H.; Tomsia, A. P.; Inai, N.; Watanabe, L. G.; Wu-Magidi, I.-C.; Marshall, Jr. G. W. *Acta Mater.* 1998, *46*, 2529–2539.
17. Seth, B. R. Transition Theory of Elastic–Plastic Deformation, Creep and Relaxation. *Nature* 1962, *195*, 896–897.
18. Seth, B. R. Measure Concept in Mechanics. *Int. J. Non-linear Mech.* 1966, *1* (1), 35–40.
19. Seth, B. R. Elastic–Plastic Transition in Shells and Tubes under Pressure. *Z. angew. Math. Mech.* 1963, *43*, 345–351.
20. Shahi, S.; Singh, S. B.; Thakur, P. Modeling Creep Parameter in Rotating Discs with Rigid Shaft Exhibiting Transversely Isotropic and Isotropic Material Behavior. *J. Emer. Technol. Innov. Res.* 2019, *6* (1), 387–395.
21. Sokolinokoff, I. S. *Mathematical Theory of Elasticity*. 2nd ed., McGraw-Hill: New York, 1956.

22. Thakur, P. Creep Transition Stresses of a Thick Isotropic Spherical Shell by Finitesimal Deformation Under Steady-State of Temperature and Internal Pressure. *Ther. Sci.* **2011,** *15* (2), S157–S165.

23. Thakur, P.; Shahi, S.; Gupta, N.; Singh, S. B. Effect of Mechanical Load and Thickness Profile on Creep in a Rotating Disc by using Seth's Transition Theory, AIP Conf. Proc. Amer. Inst. of Phy., USA, 2017, *1859* (1): 020024. doi.org/10.1063/1.49 90177.

24. Timoshenko, S. P.; Woinowsky-Krieger, S. *Theory of Plates and Shells.* 2nd ed., McGraw-Hill: New York, 1959.

25. You, L. H.; Zhang, J. J.; You, X. Y. Elastic Analysis of Internally Pressurized Thick-Walled Spherical Pressure Vessels of Functionally Graded Materials. *Int. J. Press. Vess. Pip.* **2005,** *82,* 347–354.

CHAPTER 4

Characterization of Material in a Solid Disc

PANKAJ THAKUR* and MONIKA SETHI

Department of Mathematics, Faculty of Science and Technology, ICFAI University, Baddi Solan 174103, Himachal Pradesh, India

Corresponding author. E-mail: pankaj_thakur15@yahoo.co.in

ABSTRACT

Seth's transition theory is applied to the problems of a solid rotating disc. By specializing the results to perfectly plastic material, the usual statically determinate stress distribution is recovered, but; however, the plastic stress at the axis becomes infinite, these stresses are not meaningful. It has been observed that the discs made of clay and concrete materials required maximum values of plastic stresses at the centre of the disc as compared to a disc made of rubber.

4.1 INTRODUCTION

The theoretical and experimental investigations on the rotating solid discs have gained widespread attention due to the great practical importance in mechanical engineering. For a better utilization of the material, it is necessary to allow variation of the effective material or thickness properties in one direction of the solid disc. Rotating discs have received a great deal of attention because of its wide use in many mechanical and electronic devices. They have extensive practical engineering applications such as in steam and gas turbines, turbo generators, flywheel of internal combustion engines, turbojet engines, reciprocating engines, centrifugal compressors, and brake discs. The problems of rotating solid discs have been analyzed under various interesting assumptions and the topic is discussed in most of the standard text books on elasticity and plasticity.[1–5] Most of the research works discussed are the analytical solutions of rotating disc with simple cross section geometries of uniform thickness and of specifically variable thickness. The solution of a rotating solid disc with constant thickness was obtained by Gamer[6,7] taking into account the linear strain hardening material behavior. Gamer[6] found the elastic–plastic deformation of the rotating solid disk under the assumptions of Tresca's yield condition, its associated flow rule, and linear strain hardening. To obtain the stress distribution, they matched the plastic stresses at the same radius $r = z$ of the disc. Seth's transition theory[8] includes classical macroscopic-solving problems in elasticity, plasticity, creep, and relaxation and assumes semiempirical yield conditions. Nonlinear transition region through which yielding occurs is neglected. The transition theory is used in solving problems of generalized strain measure and asymptotic solution at critical points of the differential equations defining the deforming field. It has been successfully applied to a large number of the problems.[10–15]

In the subsequent sections of this chapter, we investigate the problem of infinitesimal deformation in a solid disc using Seth's transition theory. Numerical results have also been obtained and depicted graphically.

4.2   GOVERNING EQUATIONS

Consider a state of plane stress in the case of a solid disc having constant density with external radius b. The disc is rotating with angular velocity ω about an axis perpendicular to its plane and passing through the centre. The thickness of the disc is assumed to be constant and is taken to be sufficiently small so that the disc is effectively in a state of plane stress that is, the axial stress T_{zz} is zero.

4.2.1   DISPLACEMENT COORDINATES

For this problem, displacement components in cylindrical polar coordinates are given as[9]:

$$u = r\,(1-\beta);\ v = 0;\ w = dz, \tag{4.1}$$

where β is position function, depending on the value of $r = \sqrt{x^2 + y^2}$ only, and d be a constant.

The infinitesimal strain components are given[12] as:

$$\overset{A}{e_{rr}} \equiv \frac{\partial u}{\partial r} = \left[1 - (r\beta' + \beta)\right],\ \overset{A}{e_{\theta\theta}} \equiv \frac{u}{r} = \left[1 - \beta\right],$$

$$\overset{A}{e_{zz}} \equiv \frac{\partial w}{\partial z} = d,\ \overset{A}{e_{r\theta}} = \overset{A}{e_{\theta z}} = \overset{A}{e_{zr}} = 0 \tag{4.2}$$

where $\beta' = d\beta\,/\,dr$.

4.2.2 GENERALIZED STRAIN COMPONENTS

The generalized components of strain are given[8]:

$$e_{rr} = \frac{1}{n}\left[1 - \{2(r\beta' + \beta) - 1\}^{n/2}\right],\ e_{\theta\theta} = \frac{1}{n}\left[1 - \{2\beta - 1\}^{n/2}\right],$$

$$e_{zz} = \frac{1}{n}\left[1 - (1 - 2d)^{n/2}\right],\ e_{r\theta} = e_{r\theta} = e_{zr} = 0 \tag{4.3}$$

where $\beta' = d\beta\,/\,dr$.

4.2.3 STRESS–STRAIN RELATION

The stress-strain relations for isotropic material become are given[1] as:

$$T_{ij} = \lambda\delta_{ij}I_1 + 2\mu e_{ij},\ (i, j = 1, 2, 3) \tag{4.4}$$

where T_{ij} is the stress components e_{ij} strain components λ, μ are Lame's constants $I_1 = e_{kk}$ is the first strain invariant and δ_{ij} is the Kronecker's delta. Equation (4.4) for this problem becomes

$$T_{rr} = \frac{2\lambda\mu}{\lambda + 2\mu}\left[e_{rr} + e_{\theta\theta}\right] + 2\mu e_{rr},\ T_{\theta\theta} = \frac{2\lambda\mu}{\lambda + 2\mu}\left[e_{rr} + e_{\theta\theta}\right] + 2\mu e_{\theta\theta} - \frac{2\mu\xi\theta}{\lambda + 2\mu},$$

$$T_{zz} = T_{zr} = T_{r\theta} = T_{\theta z} = 0 \tag{4.5}$$

Substituting eq (4.2) in eq (4.4), the strain components in terms of stresses are obtained as:

$$e_{rr} = \frac{\partial u}{\partial r} = \frac{1}{E}[T_{rr} - vT_{\theta\theta}] = \frac{1}{E}\left[T_{rr} - \left(\frac{1-C}{2-C}\right)T_{\theta\theta}\right],$$

$$e_{\theta\theta} = \frac{u}{r} = \frac{1}{E}[T_{\theta\theta} - vT_{rr}] = \frac{1}{E}\left[T_{\theta\theta} - \left(\frac{1-C}{2-C}\right)T_{rr}\right],$$

$$e_{zz} = -\frac{v}{E}[T_{rr} - T_{\theta\theta}] = -\left(\frac{1-C}{2-C}\right)\frac{1}{E}[T_{rr} - T_{\theta\theta}] \tag{4.6}$$

where $E = \frac{\mu(3\lambda + 2\mu)}{(\lambda + \mu)}$ and $v = \frac{\lambda}{2(\lambda + \mu)}$ and $v = \left(\frac{1-C}{2-C}\right)$. Substituting

eq (4.3) in eq (4.5), the stresses are obtained as:

$$T_{rr} = \frac{2\mu}{n}\left[3 - 2C - \{2\beta(P+1)-1\}^{n/2}(2-C) - (2\beta-1)^{n/2}(1-C)\right]$$

$$T_{\theta\theta} = \frac{2\mu}{n}\left[3 - 2C - \{2\beta(P+1)-1\}^{n/2}(1-C) - (2\beta-1)^{n/2}(2-C)\right]$$

$$T_{zz} = T_{zr} = T_{r\theta} = T_{\theta z} = 0 \tag{4.7}$$

where $r\beta' = \beta P$ and $C = \frac{2\mu}{\lambda + 2\mu}$ be the compressibility factor in terms λ of and μ terms.

4.2.4 EUATION OF EQUILIBRIUM

The equations of equilibrium are all satisfied except

$$\frac{d}{dr}(rT_{rr}) - T_{\theta\theta} + \rho\omega^2 r^2 = 0 \tag{4.8}$$

where T_{rr} be the radial stress; $T_{\theta\theta}$ circumferential stresses and ρ be the constant density of the solid disc.

4.2.5   ASYMPTOTIC SOLUTION AT TRANSITION POINTS

Using eq (4.7) in eq (4.8), we get a nonlinear differential equation in β as:

$$(2-C)n\beta^2 P\{2\beta(P+1)-1\}^{\left(\frac{n}{2}-1\right)}\frac{dP}{d\beta}$$

$$=\left|\begin{array}{c}\dfrac{n\rho\omega^2 r^2}{2\mu}-\{2\beta(P+1)-1\}^{n/2}\\[2mm]\left[1+\right.\\[1mm]\left.\dfrac{n\beta P(P+1)(2-C)}{\{2\beta(P+1)-1\}}\right]+\{2\beta-1\}^{n/2}\left[1-\dfrac{n\beta P(1-C)}{\{2\beta-1\}}\right]\end{array}\right| \qquad (4.9)$$

4.2.6   CRITICAL OR TRANSITION POINTS

Transition points of β in eq (4.9) are $P \to 0$ and $P \to \pm\infty$. However $P \to 0$ gives nothing of importance.

4.2.7   BOUNDARY CONDITION

The boundary conditions are:

$$u = 0 \ at \ r = 0 \ and \ T_{rr} = 0 \ at \ r = b \qquad (4.10)$$

4.3   SOLUTION

It has been shown[8–12] that the asymptotic solution through the principal stress leads from elastic state to plastic state at the transition point $P \to +\infty$. We define the transition function R_1 as:

$$R_1 = \frac{n}{2\mu}[T_{\theta\theta}] = \left[(3-2C)-\{2\beta(P+1)-1\}^{n/2}(1-C)-(2\beta-1)^{n/2}(2-C)\right] \qquad (4.11)$$

Taking the logarithmic differentiation of eq (4.11) with respect to r and using eq (4.9), we get:

$$\frac{d\left(\log R_1\right)}{dr} = -\frac{\left(\dfrac{1-C}{2-C}\right)\left[\begin{array}{c}\dfrac{n\rho\omega^2 r^2}{2\mu\beta^n}-\{2\beta(P+1)\}^{n/2} \\[2mm] +(2\beta-1)^{n/2}-n\beta P(1-C)(2\beta-1)^{n/2}\end{array}\right]}{r\left[\begin{array}{c}+n\beta P(2-C)(2\beta-1)^{n/2} \\[2mm] 3-2C-\{2\beta(P+1)-1\}^{n/2}(1-C) \\[2mm] -(2\beta-1)^{n/2}(2-C)\end{array}\right]} \qquad (4.12)$$

Taking the asymptotic value of eq (4.12) as $P \to +\infty$ and after integration we get:

$$R_1 = A_1 r^{-\frac{1}{2-C}} \qquad (4.13)$$

where A_1 is a constant of integration which can be determined by boundary condition. From eq (4.11) and (4.13), we have:

$$T_{\theta\theta} = \left(\frac{2\mu}{n}\right)A_1 r^{-\frac{1}{2-C}} \qquad (4.14)$$

Substituting eq (4.14) in eq (4.8) and integrating, we get:

$$T_{rr} = \left\{\frac{2\mu(2-C)}{n(1-C)}\right\}A_1 r^{-\frac{1}{2-C}} - \frac{\rho\omega^2 r^2}{3} + \frac{B_1}{r}. \qquad (4.15)$$

where B_1 is a constant of integration which can be determined by boundary condition. Substituting eqs (4.14) and (4.15) in eq (4.6), we get:

$$\frac{\partial u}{\partial r} = \frac{1}{E}\left[\left(\frac{2\mu}{n}\right)A_1 r^{-\frac{1}{2-C}}\left\{\frac{3-2C}{(1-C)(2-C)}\right\} - \frac{\rho\omega^2 r^2}{3} + \frac{B_1}{r}\right] \qquad (4.16)$$

$$\frac{u}{r} = \frac{(1-C)}{E(2-C)}\left[\frac{\rho\omega^2 r^2}{3} - \frac{B_1}{r}\right], \qquad (4.17)$$

where $E = \dfrac{2\mu(3-2C)}{(2-C)}$ is the Young's modulus. Integrating eq (4.16) with respect to r, we get:

$$u = \frac{1}{E}\left[\left(\frac{2\mu}{n}\right)A_1 r^{\frac{1-C}{2-C}}\left\{\frac{(3-2C)}{(1-C)^2}\right\} - \frac{\rho\omega^2 r^3}{9} + B_1 \log r\right] + D. \qquad (4.18)$$

where D is a constant of integration, which can be determined by boundary condition. Comparing eqs (4.16) and (4.18), we get:

$$\left(\frac{2\mu}{n}\right)A_1 r^{\frac{1-C}{2-C}}\left\{\frac{3-2C}{(1-C)^2}\right\} = \left[\begin{array}{c}\dfrac{\rho\omega^2 r^3}{9}\left(\dfrac{5-4C}{2-C}\right) \\ -B_1\left\{\dfrac{(1-C)+(2-C)\log r}{(2-C)}\right\} - DE\end{array}\right] \qquad (4.19)$$

$$\text{and } u = \frac{(1-C)}{E(2-C)}\left[\frac{\rho\omega^2 r^3}{3} - B_1\right] \qquad (4.20)$$

Using boundary conditions (4.10) in eq (4.20), we get $B_1 = 0$. Using eq (4.19) in eq (4.15) and using boundary condition (4.10), we get:

$$D = \frac{1}{E}\left\{\frac{\rho\omega^2 b^3}{9}\left(\frac{5-4C}{2-C}\right) - \frac{(3-2C)}{(1-C)(2-C)}\frac{\rho\omega^2 b^3}{3}\right\} \qquad (4.21)$$

Putting values of B, D and using eq (4.19) in eqs (4.14) and (4.15) respectively, we get the plastic stresses and displacement as:

$$rT_{rr} = \frac{\rho\omega^2}{9}\left(r^3 - b^3\right)\left(\frac{C^2 - 3C - 5}{3 - 2C}\right),$$

$$rT_{\theta\theta} = \frac{\rho\omega^2}{3(3-2C)}\left(\frac{1-C}{2-C}\right)\left[(5-4C)(1-C)r^3 - \left(4C^2 - 3C - 4\right)b^3\right]$$

$$\text{and } u = \frac{(1-C)}{E(2-C)}\frac{\rho\omega^2 r^3}{3}. \qquad (4.22)$$

At the centre the stress for a solid disc is given when $r = 0$. With radius equal to zero the above eq (4.22) produced infinite stresses whatever the speed of rotation. These stresses can be neglected. Stresses and displacement for fully plastic state $(C \to 0)$, are obtained from eqs (4.19), (4.20), (4.21), and (4.18) as:

$$rT_{rr} = \frac{5\rho\omega^2}{27}\left(b^3 - r^3\right), rT_{\theta\theta} = \frac{\rho\omega^2}{18}\left[5r^3 + 4b^3\right], \text{ and } u = \frac{1}{2E}\frac{\rho\omega^2 r^3}{3} \quad (4.23)$$

Meaning: Sigma $r = T_{rr} / \rho\omega^2 b^2$, Sigma theta $= T_{\theta\theta} / \rho\omega^2 b^2$ and Displacement $= u / \rho\omega^2 b^2$.

4.4   NUMERICAL ILLUSTRATION AND DISCUSSION

To see the effect of stresses and displacement distribution on a rotating solid disc, following values have been taken C = 0 (e.g., rubber), 0.25 (i.e., Clay material), 0.75 (Concrete material) and shown in Figure 4.1. In Figure 4.1, the curve has been drawn between $T_{rr} / \rho\omega^2 b^2$, $T_{\theta\theta} / \rho\omega^2 b^2$ and $u / \rho\omega^2 b^2$ required for yielding at the outer surface of the rotating solid disc along the radii ratio R = r/b. For the perfectly plastic material of the solid disc, usual statically determinate stresses and displacement is recovered but, since the plastic stress at this axis becomes infinite, these stresses are not meaningful. Similar results were obtained by Gamer,[6] for rotating the solid disc to account for v = 0.333 for linear strain. Clay and concrete materials increased the values of stresses at the centre of the disc as compared to the disc made of rubber material.

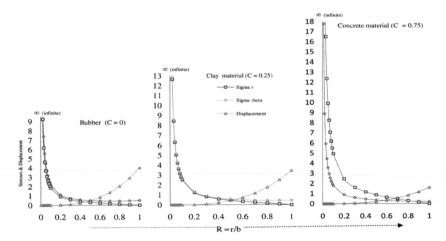

FIGURE 4.1   Graph between $T_{rr} / \rho\omega^2 b^2$, $T_{\theta\theta} / \rho\omega^2 b^2$ and $u / \rho\omega^2 b^2$ required for yielding at the outer surface of the rotating solid disc along the radii ratio R = r/b.

KEYWORDS

- **isotropic structures**
- **displacement**
- **stress concentrations**
- **disc**
- **deformation**

REFERENCES

1. Sokolnikoff, S. *Mathematical Theory of Elasticity*, 2nd ed.; McGraw Hill: New York, 1952; pp. 65–79.
2. Johnson, W.; Mellor, P. B. *Plasticity for Mechanical Engineers*; Van-Nostrand Reinhold Company: London, 1962.
3. Seth, B. R. Transition Theory of Elastic–Plastic Deformation, Creep and Relaxation. *Nature* **1962,** *195,* 896–897.
4. Seth, B. R. Measure Concept in Mechanics. *Int. J. Non-linear Mech.* **1966,** *1,* 35–40.
5. Timoshenko, S. P.; Goodier, J. N. *Theory of Elasticity.* McGraw-Hill: New York, 1970.
6. Blazynski, T. N. *Applied Elastic-Plasticity of Solids*; McMillan Press Ltd.: London, 1983.
7. Gamer, U. Elastic-plastic Deformation of the Rotating Solid Disk. *Ingenieur-Archiv* **1984,** *54,* 345–354.
8. Gamer, U. Stress Distribution in the Rotating Elastic-Plastic Disk. *ZAMM* **1985,** *65,* T136–137.
9. Chakrabarty, J. *Theory of Plasticity*; McGraw-Hill: New York, 1987.
10. Gupta, S. K.; Thakur, P. Creep Transition in a Thin Rotating Disc with Rigid Inclusion. *Defence Sci. J.* **2007,** *57* (2), 185–195.
11. Thakur, P.; Gupta, S. K. Thermo Elastic–Plastic Transition in a Thin Rotating Disc with Inclusion. *Therm. Sci.* **2007,** *11* (1), 103–118.
12. Thakur, P. Elastic–Plastic Transition Stresses in a Thin Rotating Disc with Rigid Inclusion by Infinitesimal Deformation under Steady-State Temperature. *Therm. Sci.* **2010,** *14* (1), 209–219.
13. Thakur, P.; Singh, S. B.; Kaur, J. Thermal Creep Stresses and Strain Rates in a Circular Disc with Shaft having Variable Density. *Eng. Computation* **2016,** *33* (3), 698–712.
14. Thakur, P.; Singh, S. B.; Gupta, N. Creep Strain Rates Analysis in Cylinder under Temperature Gradient Materials by using Seth's Theory. *Eng. Computation* **2017,** *34* (3), 1020–1030.
15. Thakur, P.; Sethi, M. Creep Damage Modeling in a Transversely Isotropic Rotating Disc with Load and Density Parameter. *Struct. Integr. Life* **2018,** *18* (3), 207–214.

CHAPTER 5

Characterization of Material in a Rotating Disc Subjected to Thermal Gradient

PANKAJ THAKUR* and MONIKA SETHI

Department of Mathematics, Faculty of Science and Technology, ICFAI University, Baddi Solan 174103, Himachal Pradesh, India

Corresponding author. E-mail: pankaj_thakur15@yahoo.co.in

ABSTRACT

Seth's transition theory is applied to the problems of characterization of material in a rotating disc subjected to a thermal gradient. It has been observed that the disc made of saturated clay, copper, and cast iron materials yields at the outer surface at a higher angular speed as compared to the disc made of rubber material at a steady-state temperature, whereas disc made of clay, copper, cast iron as well as rubber material yields at a lesser angular speed as compared to a rotating disc at the room temperature. With the introduction of temperature, the radial as well as hoop stresses, decrease with the increased value of the temperature at the elastic–plastic stage, but the reverse result obtained for a fully plastic case.

5.1 INTRODUCTION

Rotating disc forms an essential part of the design of rotating machinery, namely, rotors turbines, flywheel, compressors, high-speed gear engine, and so forth. The rotating disc used in machine and structural applications has generated considerable interest in many problems in the domain of mechanics of solid. Solution for thin isotropic discs can be found in most of the standard elasticity and plasticity textbooks.[1-3,6,9] Parmaksigoğlu and Güven[10] analyzed the problem of plastic stress distribution in a rotating disc with rigid inclusion under a radial temperature gradient under the assumptions of Tresca's yield condition, its associated flow rule, and linear strain hardening. To obtain the stress distribution, they matched the plastic stresses at the same radius $r = z$ of the disc. Seth's transition theory[4] includes classical macroscopic-solving problems in elasticity, plasticity, creep, and relaxation and assumes semiempirical yield conditions. Nonlinear transition region through which yielding occurs is neglected. The transition theory is used in solving problems of generalized strain measure and asymptotic solution at critical points of the differential equations defining the deforming field and has been successfully applied to a large number of the problems.[4,5,8,11-13]

5.2   MATHEMATICAL MODEL AND GOVERNING EQUATION

Consider a thin rotating disc having constant density with the central bore of radius r_i and external radius r_o. The rotating disc is mounted on rigid shaft and is rotating with an angular velocity ω about an axis perpendicular to its plane and passing through the centre. The thickness of the disc is assumed to be constant and is taken to be sufficiently small so that the disc is effectively in a state of plane stress, that is, the axial stress τ_{zz} is zero. Let a uniform temperature Θ_0 be applied at the inner surface of the rotating disc.

5.2.1   DISPLACEMENT COORDINATES

For this problem, displacement components in cylindrical polar coordinates are given as[9]:

$$u = r(1-\eta); v = 0; w = dz, \qquad (5.1)$$

where η is position function, depending on the value of $r = \sqrt{x^2 + y^2}$ only, and d be a constant.

5.2.2 GENERALIZED STRAIN COMPONENTS

The generalized strain components are given by:

$$e_{rr} = \frac{1}{n}\left[1-\left\{2\left(r\eta'+\eta\right)-1\right\}^{n/2}\right], e_{\theta\theta} = \frac{1}{n}\left[1-\left\{2\eta-1\right\}^{n/2}\right],$$

$$e_{zz} = \frac{1}{n}\left[1-\left(1-2d\right)^{n/2}\right], e_{r\theta} = e_{r\theta} = e_{zr} = 0, \tag{5.2}$$

where $\eta' = d\eta / dr$.

5.2.3 STRESS–STRAIN RELATION

The stress–strain relations for thermoplastic in an isotropic media is given by[1]:

$$\tau_{ij} = \lambda\delta_{ij}I_1 + 2\mu e_{ij} - \xi\Theta\delta_{ij}, \left(i, j = 1, 2, 3\right) \tag{5.3}$$

where e_{ij}, τ_{ij} are strain and stress tensor; $I_1 = e_{kk}$ ($k = 1,2,3$) is strain invariant; δ_{ij} is Kronecker's delta; Θ be a temperature; $\xi = \alpha(3\lambda + 2\mu)$; α being the coefficient of thermal expansion and λ, μ are Lame's constants. Further, Θ has to satisfy heat equation which gives[7]:

$$\nabla^2\Theta = 0. \tag{5.4}$$

Equation (5.3) for this problem becomes

$$\tau_{rr} = \frac{2\lambda\mu}{\lambda+2\mu}\left[e_{rr}+e_{\theta\theta}\right]+2\mu e_{rr} - \frac{2\mu\xi\Theta}{\lambda+2\mu},$$

$$\tau_{\theta\theta} = \frac{2\lambda\mu}{\lambda+2\mu}\left[e_{rr}+e_{\theta\theta}\right]+2\mu e_{\theta\theta} - \frac{2\mu\xi\Theta}{\lambda+2\mu},$$

$$\tau_{zz} = \tau_{zr} = \tau_{r\theta} = \tau_{\theta z} = 0. \tag{5.5}$$

From eq (5.3), strain components in terms of stresses are obtained as:

$$e_{rr} = \frac{1}{E}\left[\tau_{rr}-v\tau_{\theta\theta}\right]+\alpha\Theta, e_{\theta\theta} = \frac{1}{E}\left[\tau_{\theta\theta}-v\tau_{rr}\right]+\alpha\Theta,$$

$$e_{zz} = -\frac{v}{E}\left[\tau_{rr}-\tau_{\theta\theta}\right]+\alpha\Theta, e_{rr} = e_{\theta z} = e_{zr} = 0, \tag{5.6}$$

where $E = \mu(3\lambda + 2\mu)/(\lambda + \mu)$ is Young's modulus and $v = \lambda/2\,(\lambda + \mu) = 1 - c/2 - c$ be Poisson's ratio in terms of compressibility factor and Lame's constants. From eqs (5.2) and (5.5), we get the stresses as:

$$\tau_{rr} = \frac{2\mu}{n}\left[3 - 2c - \left\{2\eta\left(T+1\right)-1\right\}^{n/2}\left(2-c\right) - \left(2\eta-1\right)^{n/2}\left(1-c\right) - \frac{nc\xi\Theta}{2\mu}\right],$$

$$\tau_{\theta\theta} = \frac{2\mu}{n}\left[3 - 2c - \left\{2\eta\left(T+1\right)-1\right\}^{n/2}\left(1-c\right) - \left(2\eta-1\right)^{n/2}\left(2-c\right) - \frac{nc\xi\Theta}{2\mu}\right],$$

$$\tau_{zz} = \tau_{zr} = \tau_{r\theta} = \tau_{\theta z} = 0 \tag{5.7}$$

where $c = 2\mu/\lambda + 2\mu$ be the compressibility factor in terms of λ and μ and $r\eta' = \eta T$.

5.2.4   EQUATION OF EQUILIBRIUM

The equations of equilibrium are all satisfied except:

$$\frac{d}{dr}\left(r\tau_{rr}\right) - \tau_{\theta\theta} + \rho\omega^2 r^2 = 0. \tag{5.8}$$

where and τ_{rr} be the radial stress; $\tau_{\theta\theta}$ circumferential stresses and ρ be the constant density of the rotating disc.

5.2.5   ASYMPTOTIC SOLUTION AT TRANSITION POINTS

Using eqs (5.7) and (5.12) in eq (5.8), we get a nonlinear differential equation in η given as:

$$(2-c)\,n\eta^2 T\left\{2\eta\left(T+1\right)-1\right\}^{\left(\frac{n}{2}-1\right)}\frac{dT}{d\eta}$$

$$= \left\{\begin{array}{l}\left[\begin{array}{l}\left(\dfrac{n\rho\omega^2 r^2}{2\mu} - \left\{2\eta\left(T+1\right)-1\right\}^{n/2}\left[1 + \dfrac{n\eta T\left(T+1\right)\left(2-c\right)}{\left\{2\eta\left(T+1\right)-1\right\}}\right]\right) \\[4mm] + \left\{2\eta-1\right\}^{n/2}\left[1 - \dfrac{n\eta T\left(1-c\right)}{\left\{2\eta-1\right\}}\right]\end{array}\right] - \dfrac{nc\xi\overline{\Theta}_0}{2\mu}\end{array}\right\} \tag{5.9}$$

where $\bar{\Theta}_0 = \dfrac{\Theta_0}{\log(a/b)}$.

5.2.6 CRITICAL OR TRANSITION POINTS

Transition points of η in eq (5.9) are $T \to 0$ and $T \to +\infty$. At the transition point $T \to 0$ is nothing of importance.

5.2.7 BOUNDARY CONDITION

The rotating disc considered in the present study is subjected to a temperature gradient field and infinitesimal deformation. The inner surface of the disc is assumed to be fixed to a shaft so that the isothermal conditions prevail on it. The inner surface of the disc is applied uniform temperature gradient. Thus, the boundary conditions of the problem are:

$$r = r_i, \ u = 0; r = b, \ \tau_{rr} = 0 \ at \ r = r_o \tag{5.10}$$

where u and T_{rr} denote displacement, stress along the radial direction applied at the external surface. The temperature field satisfying eq (5.4) and

$$\Theta = \Theta_0 \ at \ r = r_i, \Theta = 0 \ at \ r = r_o \tag{5.11}$$

where Θ_0 is constant, is given by

$$\Theta = \frac{\Theta_0 \log(r/r_o)}{\log(r_i/r_0)} \tag{5.12}$$

5.3 SOLUTION

It has been shown[4,5,8,11–13] that the asymptotic solution through the principal stress leads from elastic state to plastic state at the transition point $T \to +\infty$. We define the transition function Z as:

$$Z = \frac{n}{2\mu}[\tau_{\theta\theta} + c\xi\Theta] = \left[(3-2c) - \{2\eta(T+1)-1\}^{n/2}(1-c) - (2\eta-1)^{n/2}(2-c)\right] \tag{5.13}$$

By taking the logarithmic differentiation of eq (5.13) with respect to r and using eq (5.9), we get:

$$\frac{d(\ln Z)}{dr} = -\frac{\left[\left(\frac{1-c}{2-c}\right)\left[\frac{\frac{n\rho\omega^2 r^2}{2\mu\beta^n} - \{2\eta(T+1)\}^{n/2} + (2\eta-1)^{n/2}}{-n\eta T(1-c)(2\eta-1)^{n/2} - \frac{nc\xi\overline{\Theta}_0}{2\mu\eta^n}}\right] + n\eta T(2-c)(2\eta-1)^{n/2}\right]}{r\left[3-2c-\{2\eta(T+1)-1\}^{n/2}(1-c)-(2\eta-1)^{n/2}(2-c)\right]} \tag{5.14}$$

The asymptotic value from eq (5.14) as $T \to +\infty$, and integrating we get:

$$Z = Lr^{-1/(2-c)} \tag{5.15}$$

where L is a constant of integration. From eqs (5.13) and (5.15), we have:

$$\tau_{\theta\theta} = \left(\frac{2\mu}{n}\right)Lr^{-1/(2-c)} - \frac{c\xi\Theta_0 \log(r/r_o)}{\log(r_i/r_o)}. \tag{5.16}$$

Using eq (5.16) into eq (5.8) and integrating, we get:

$$\tau_{rr} = \left\{\frac{2\mu(2-c)}{n(1-c)}\right\}Lr^{-1/(2-c)} - \frac{c\xi\Theta_0 \log(r/r_o)}{\ln(r_i/r_o)} + \frac{c\xi\Theta_0}{\ln(r_i/r_o)} - \frac{\rho\omega^2 r^2}{3} + \frac{M}{r} \tag{5.17}$$

where M is a constant of integration. Substituting eqs (5.16) and (5.17) in eq (5.6), we get:

$$\frac{\partial u}{\partial r} = \frac{1}{E}\left[\left(\frac{2\mu}{n}\right)Lr^{-1/2-c}\left\{\frac{3-2c}{(1-c)(2-c)}\right\} + \frac{\alpha E\theta_0(2-c)}{\ln(r_i/r_o)} - \frac{\rho\omega^2 r^2}{3} + \frac{M}{r}\right], \tag{5.18}$$

$$\frac{u}{r} = \frac{(1-c)}{E(2-c)}\left[\frac{\rho\omega^2 r^2}{3} - \frac{\alpha E\Theta_0(2-c)}{\ln(r_i/r_o)} - \frac{M}{r}\right], \tag{5.19}$$

where $c\xi = 2\mu\alpha(3-2c)$ and $E = 2\mu(3-2c)/(2-c)$ is the Young's modulus. By integrating equation (5.18) with respect to r, we get:

$$u = \frac{1}{E}\left[\left(\frac{2\mu}{n}\right)Lr^{\frac{1-c}{2-c}}\left\{\frac{(3-2c)}{(1-c)^2}\right\} + \frac{\alpha E\Theta_0(2-C)r}{\ln(r_i/r_o)} - \frac{\rho\omega^2 r^3}{9} + M\log r\right] + N \tag{5.20}$$

where D is a constant of integration. From eqs (5.19) and (5.20), we get:

$$\left(\frac{2\mu}{n}\right)Lr^{\frac{1-c}{2-c}}\left\{\frac{3-2c}{(1-c)^2}\right\}$$

$$=\left[\frac{\rho\omega^2 r^3}{9}\left(\frac{5-4c}{2-c}\right)-\frac{\alpha E\Theta_0 r(3-2c)}{\ln(r_i/r_o)}-M\left\{\frac{(1-c)+(2-c)\ln r}{(2-c)}\right\}-NE\right] \quad (5.21)$$

and $u=\dfrac{(1-c)}{E(2-c)}\left[\dfrac{\rho\omega^2 r^3}{3}-\dfrac{\alpha E\Theta_0(2-c)r}{\ln(r_i/r_o)}-M\right]$ \qquad (5.22)

Using boundary conditions (5.10) in eq (5.22), we get:

$$M=\frac{\rho\omega^2 r_i^3}{3}-\frac{\alpha E\Theta_0 r_i(2-c)}{\ln(r_i/r_o)} \quad (5.23)$$

Substituting eq (5.21) into eq (5.17) and using boundary condition from (5.10) and eq (5.13), we get:

$$N=\frac{1}{E}\left\{\begin{array}{l}\left[\dfrac{\rho\omega^2 r_o^3}{9}\left(\dfrac{5-4c}{2-c}\right)-\dfrac{\alpha E\Theta_0 r_o(3-2c)}{\ln(r_i/r_o)}-\left(\dfrac{\rho\omega^2 r_i^3}{3}-\dfrac{\alpha E\Theta_0 r_i(2-c)}{\ln(r_i/r_o)}\right)\left[\left(\dfrac{1-c}{2-c}\right)+\ln r_o\right]\right] \\ +\dfrac{(3-2c)}{(1-c)(2-c)}\left[\dfrac{\rho\omega^2\left(r_i^3-r_o^3\right)}{3}+\dfrac{\alpha E\Theta_0(2-C)(r_o-r_i)}{\ln(r_i/r_o)}\right]\end{array}\right\} \quad (5.24)$$

By using eqs (5.21), (5.23), and (5.24) in eqs (5.16) and (5.17) respectively, we get:

$$\tau_{\theta\theta}=\left\{\begin{array}{l}\dfrac{\rho\omega^2}{3r}\left[\dfrac{1}{3}\left(\dfrac{5-4c}{2-c}\right)\dfrac{(1-c)^2\left(r^3-r_o^3\right)}{(3-2c)}-r_i^3\ln\left(\dfrac{r}{r_o}\right)\dfrac{(1-c)^2}{(3-2c)}+\left(\dfrac{1-c}{2-c}\right)\left(r_o^3-r_i^3\right)\right] \\ -\dfrac{\alpha E\Theta_0}{\ln(r_i/r_o)}\left[\begin{array}{l}\dfrac{(1-c)^2(r-r_o)}{r}-\dfrac{r_i}{r}\ln\left(\dfrac{r}{r_o}\right)\dfrac{(2-c)(1-c)^2}{(3-2c)}+ \\ \dfrac{(r_o-r_i)(1-c)}{r}+(2-c)\ln\left(\dfrac{r}{r_o}\right)\end{array}\right]\end{array}\right\},$$

$$\tau_{rr}=\left\{\begin{array}{l}\dfrac{\rho\omega^2}{3r}\left[\dfrac{1}{3}\left(\dfrac{5-4c}{3-2c}\right)(1-c)\left(r^3-r_o^3\right)-r_i^3\ln\left(\dfrac{r}{r_o}\right)\dfrac{(1-c)(2-c)}{(3-2c)}+r_o^3-r^3\right] \\ -\dfrac{\alpha E\Theta_0(2-c)}{\ln(r_i/r_o)}\left[c\left(\dfrac{r_o}{r}-1\right)+\ln\left(\dfrac{r}{r_o}\right)-\dfrac{r_i}{3r(2-c)}\ln\left(\dfrac{r}{r_o}\right)\right]\end{array}\right\},$$

and
$$u = \frac{(1-c)}{(2-c)}\left[\frac{\rho\omega^2}{3}\left(r^3 - r_i^3\right) - \frac{\alpha E\Theta_0(2-c)}{\ln(r_i/r_o)}(r - r_i)\right]$$ (5.25)

Yielding at the initial stage: It has been seen from eq (5.25) that $|\tau_{\theta\theta}|$ is maximum at the outer surface (that is, at $r = r_o$), therefore, yielding will take place at the outer surface of the rotating disc and eq (5.15) becomes

$$|\tau_{\theta\theta}|_{r=r_o} = \left|\frac{\rho\omega^2\left(r_o^3 - r_i^3\right)}{3r_o}\left(\frac{1-c}{2-c}\right) + \alpha E\theta_0\left[\frac{(1-c)(r_o - r_i)}{r_o\ln(r_o/r_i)}\right]\right| = Y(say).$$

The angular velocity ω_i necessary for the initial yielding stage is given by

$$\Omega_i^2 = \frac{\rho\omega_i^2 r_o^2}{Y} = \left\{\frac{3(2-c)}{\left(1 - r_i^3/r_o^3\right)(1-c)} - 3(2-c)\left(\frac{\alpha E\Theta_0}{Y}\right)\left[\frac{(1 - r_i/r_o)}{\left(1 - r_i^3/r_o^3\right)\ln(r_o/r_i)}\right]\right\}$$ (5.26)

and $\omega_i = \dfrac{\Omega_i}{r_o}\sqrt{\dfrac{Y}{\rho}}.$

Fully plastic stage: The angular velocity ω_f for which the disc becomes fully plastic sate $(c \to 0)$ at $r = r_i$ is given by eq (5.25) as:

$$\Omega_f^2 = \frac{\rho\omega_f^2 r_o^2}{Y^*} =$$

$$\left[4\left(\frac{r_i}{r_o}\right)\left(\frac{\alpha E\Theta_0}{Y^*}\right)\frac{1}{\left\{\frac{2}{9}\left(1 - \frac{r_i^3}{r_o^3}\right) - \frac{1}{3}\left(\frac{r_i}{r_o}\right)^3\ln\left(\frac{r_i}{r_o}\right)\right\}} + 3\left(\frac{r_i}{r_o}\right)\frac{1}{\left\{\frac{2}{9}\left(1 - \frac{r_i^3}{r_o^3}\right) - \frac{1}{3}\left(\frac{r_i}{r_o}\right)^3\ln\left(\frac{r_i}{r_o}\right)\right\}}\right],$$ (5.27)

and $\omega_f = \dfrac{\Omega_f}{r_o}\sqrt{\dfrac{Y^*}{\rho}}.$

Nondimensional components: We introduce the following nondimensional components as: $R = r/r_o$; $R_0 = r_i/r_o$; $\sigma_r = \tau_{rr}/Y$; $\sigma_\theta = \tau_{\theta\theta}/Y$; $\Theta_1 = aE\,\Theta_0/Y$; and $\bar{u} = uE/Yr_o$.

  Stresses, displacement, and angular speed at the initial stage: The elastic–plastic stresses, angular velocity, and displacement from eqs (5.25) and (5.26) in nondimensional form becomes

$$\sigma_\theta = \left\{ \begin{array}{l} \dfrac{\Omega_i^2}{3R}\left[\dfrac{1}{3}\left(\dfrac{5-4c}{2-c} \right)\dfrac{(1-c)^2}{(3-2c)}\left(R^3 -1 \right) - R_0^3 \ln R\dfrac{(1-c)^2}{(3-2c)} + \left(\dfrac{1-c}{2-c} \right)\left(1-R_0^3 \right) \right] \\[4mm] -\dfrac{\Theta_1}{\ln R_0}\left[\dfrac{(1-C)^2(R-1)}{R} + \dfrac{R_0}{R}\ln R\dfrac{(2-c)(1-c)^2}{(3-2c)} + \dfrac{(1-R_0)(1-c)}{R} + (2-c)\ln R \right] \end{array} \right\},$$

$$\sigma_r = \left\{ \begin{array}{l} \dfrac{\Omega_i^2}{3R}\left[\dfrac{1}{3}\left(\dfrac{5-4c}{3-2c} \right)(1-C)\left(R^3 -1 \right) - R_0^3 \log R\dfrac{(1-c)(2-c)}{(3-2c)} + 1 - R^3 \right] - \dfrac{\theta_1(2-c)}{\ln R_0}\left[\begin{array}{l} c\dfrac{(1-R)}{R} \\[2mm] +\ln R - \dfrac{R_0 \ln R}{3R(2-c)} \end{array} \right] \end{array} \right\},$$

and $\quad \overline{u} = \left(\dfrac{1-c}{2-c} \right)\left[\dfrac{\Omega_i^2}{3}\left(R^3 - R_0^3 \right) - \dfrac{\Theta_1(2-c)}{\log R_0}(R-R_0) \right],$ (5.28)

and $\quad \Omega_i^2 = \dfrac{3(2-c)}{(1-R_0^3)(1-c)} - \dfrac{3\Theta_1(2-c)}{(1-R_0^3)}\dfrac{(1-R_0)}{\ln(1/R_0)}$ (5.29)

Stresses, displacement, and angular speed at fully plastic stage: The stresses, displacement, and angular speed for the fully plastic state ($c \to 0$), are obtained from eqs (5.28) and (5.27) as:

$$\sigma_r = \dfrac{\Omega_f^2}{3R}\left[\dfrac{5}{9}\left(R^3 -1 \right) - \dfrac{2}{3}R_0^3 \ln R + 1 - R^3 \right] - \dfrac{2\Theta_1^*}{\ln R_0}\left[\log R - \dfrac{R_0 \ln R}{6R} \right],$$

$$\sigma_\theta = \left\{ \dfrac{\Omega_f^2}{3R}\left[\begin{array}{l} \dfrac{5}{18}\left(R^3 -1 \right) \\[3mm] -\dfrac{R_0^3 \ln R}{3} + \dfrac{1}{2}\left(1-R_0^3 \right) \end{array} \right] - \dfrac{\Theta_1^*}{\ln R_0}\left[\dfrac{(R-1)}{R} + \dfrac{2}{3}\dfrac{R_0}{R}\ln R + \dfrac{(1-R_0)}{R} + 2\ln R \right] \right\}$$

$$\overline{u}_f = \dfrac{\Omega_f^2}{6}\left(R^3 - R_0^3 \right) - \dfrac{\Theta_1^*}{\ln R_0}(R-R_0)$$ (5.30)

and $\quad \Omega_f^2 = \dfrac{3R_0}{\left[\dfrac{2}{9}\left(1-R_0^3 \right) - \dfrac{R_0^3 \ln R_0}{3} \right]} + \dfrac{4R_0\Theta_1^*}{\left[\dfrac{2}{9}\left(1-R_0^3 \right) - \dfrac{R_0^3 \ln R_0}{3} \right]},$ (5.31)

where $\quad \Theta_1^* = \dfrac{\alpha E\Theta_0}{Y^*}; \overline{u}_f = \dfrac{uE}{Y^*b}.$

5.4 NUMERICAL ILLUSTRATION AND DISCUSSION

For calculating stresses, strain rates based on the above-mentioned analysis, the following values have been taken $v = 0.5(c = 0$, incompressible material, i.e., rubber), $v = 0.42857$ ($c = 0.25$, compressible material, i.e., saturated clay), $v = 0.33$ ($c = 0.5$, compressible materials, i.e., copper), and $v = 0.21$ ($c = 0.75$, compressible materials, i.e., cast iron) and temperature $\Theta_1 = 0, 0.3, 0.45$, and 0.85 respectively.

Curves are produced in Figure 5.1, between angular speeds along with the radio ratio $R_0 = r_i/r_o$ at the initial yielding stage $R_0 = r_i/r_o$. It has been seen that the rotating disc made of compressible materials (say, saturated clay, copper, and cast iron) and of smaller radii ratio yields at the inner surface required higher angular speed as compared to disc made of incompressible material (say, rubber) at the room temperature. With the effect of thermal, disc yields at the external surface at a lesser angular speed as compared to the rotating disc at room temperature. Curves are produced, as shown in Figure 5.2, between angular speed and various radii ratio $R_0 = a/b$ for fully plastic. The disc of smaller radii ratio required higher angular speed to become fully plastic in comparison to rotating disc of higher thickness ratio and the angular speed increases with the increase in temperature.

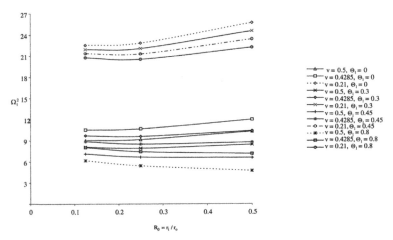

FIGURE 5.1   Graph between angular speed along the radii ration $R_0 = r_i/r_o$ at the initial yielding stage.

In Figures 5.3 and 5.4, curves have been drawn between stresses and radii ratio $R = r/r_o$ at elastic–plastic transition state and fully plastic state. It has been observed that the radial stresses are maximum at the inner surface. With the introduction of temperature, the radial as well, as hoop stresses decrease with the increased value of the temperature at the elastic–plastic stage, but the reverse result obtained for a fully plastic case.

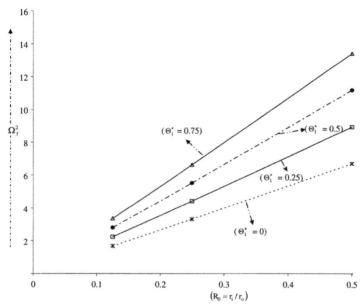

FIGURE 5.2 Graph between angular speed along the radii ratio $R_0 = r_i/r_o$ for the fully plastic stage.

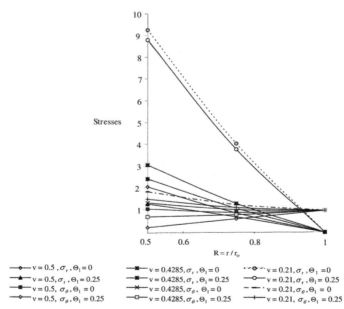

FIGURE 5.3 Graph between angular speed along the radii ratio $R_0 = r_i/r_o$ for fully plastic. stage.

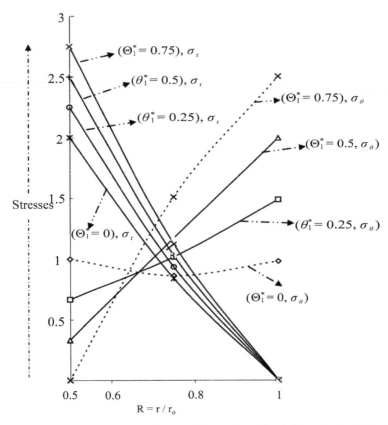

FIGURE 5.4 Graph between angular speed along the radii ratio $R_0 = r/r_0$ for fully plastic stage.

KEYWORDS

- **isotropic structures**
- **displacement**
- **stress concentrations**
- **disc**
- **infinitesimal deformation**

REFERENCES

1. Sokolnikoff, I. S. *Mathematical Theory of Elasticity*, 2nd ed; McGraw-Hill: New York, 1952; pp. 65–79.
2. Heyman, J. Plastic Design of Rotating Discs. *Proc. Inst. Mech. Engrs.* **1958**, *172*, 531–546.
3. Johnson, W.; Mellor, P. B. *Plasticity for Mechanical Engineers*; Van-Nostrand Reinhold Company: London, 1962.
4. Seth, B. R. Transition Theory of Elastic–Plastic Deformation, Creep and Relaxation. *Nature* **1962**, *195*, 896–897.
5. Seth, B. R. Measure Concept in Mechanics. *Int. J. Non-linear Mech.* **1966**, *1*, 35–40.
6. Timoshenko, S. P.; Goodier, J. N. *Theory of Elasticity*; McGraw-Hill: New York, 1970.
7. Parkus, H. Thermo-Elasticity; Springer-Verlag Wien: New York, 1976.
8. Gupta, S. K.; Dharmani, R. L. Creep Transition in Bending of Rectangular Plates. *Int. J. Non-linear Mech.* **1980**, *15* (2), 147–154.
9. Chakrabarty, J. *Theory of Plasticity*; McGraw-Hill: New York, 1987.
10. Parmaksigoğlu, C.; Güven, U. Plastic Stress Distribution in a Rotating Disc with Rigid Inclusion under a Radial Temperature Gradient. *Mech. Stru. Mach.* **1998**, *26* (1), 77 9–20.
11. Gupta, S. K.; Sharma, S.; Pathak, S. Elastic–Plastic Transition in a Thin Rotating Disc of Variable Thickness with Edge Load. *Ganita* **1998**, *49* (1), 61–75.
12. Gupta, S. K.; Pathak, S. Elastic–Plastic Transition in a Thin Rotating Disc of Variable Density with Edge Load. *Proc. Nat. Acad. Sci., India* **2000**, *70* (A), Part I, 75–86.
13. Thakur, P.; Singh, S. B.; Kaur, J. Thermal Creep Stresses and Strain Rates in a Circular Disc with Shaft having Variable Density. *Eng. Comput.* **2016**, *33* (3), 698–712.

CHAPTER 6

Optimization of Cutting Parameters for Hard Machining on Inconel 718 Using Signal to Noise Ratio and Gray Relational Analysis

IVAN SUNIT ROUT[1*], P. PAL PANDIAN[1], G. RAVICHANDRAN[1*], N. SANTHOSH[1], and G. S. HEBBAR[1]

[1]Department of Mechanical Engineering, Faculty of Engineering, CHRIST (Deemed to be University), Bangalore 560074, India

*Corresponding author. E-mail: ivan.rout@christuniversity.in; ravig_s@rediffmail.com

ABSTRACT

Inconel 718 is hard-to-cut material which is in high demand for a lot of industries including aerospace, defense, energy production, biomechanical, and chemical. Many cutting tools have been used to machine this hard material but have failed. However, ceramics and tungsten carbide have shown better results and are being used to improve the machinability and provide high cutting conditions. The aim of this paper is to optimize the cutting parameters: Spindle speed, tool feed, and depth of cut and response parameters: Surface roughness, tool wear needed for machining this hard material using signal to noise ratio (S/N), and gray relational analysis (GRA) which are considered as effective tools.

6.1 INTRODUCTION

6.1.1 INCONEL 718

It is considered as one of the hardest material to cut due to its application in extreme conditions. It has high corrosion resistance and wear resistance. It can be used at elevated temperatures of 704°C. It is a superalloy comprising of nickel 50-55%, chromium 17%, and the rest other metals such as cobalt, iron, and so on. This material is used in high compressor turbines, cryogenic tanks, nuclear fuel element spaces, tooling of hot extrusion, jet engines, fuel pumps, and high strength riveting. The machinability of this material is extremely poor due to built-up edge, high-hardening, dynamic cutting forces.[1] Hence the need for proper tools comes into effect for machining these materials.

6.1.2 CERAMICS

It is one of the hardest tools being used to machine Inconel 718 for its high toughness and high tool rigidity to withstand high temperatures during cutting operation. It can cut this hard material with ease providing better surface topography and without much wear behavior on its rake surface.[2] It is a round insert of RPGX 1204-GH of grade AS20 as shown in Figure 6.1.[6]

FIGURE 6.1 A ceramic tool insert.

6.1.3 TUNGSTEN CARBIDE

It is another economic tool being extensively used for the machining of Inconel 718. It has high wear resistance, high stability, and can be used at high temperatures during machining operations.[2] It is a round insert of RYMX 1004-ML of grade TT3540 as shown in the Figure 6.2.[6]

FIGURE 6.2   A tungsten carbide tool insert.

6.2   METHODOLOGY

6.2.1   CNC MILLING MACHINE

It is one of the most advanced conventional machines being used. It has high accuracy and provides better dimensional tolerances than other tools. It can provide better cutting speed, accurate feed rate, and suitable depth of cut during the machining operation.[3] A vertical milling machine (shown in the Fig. 6.3) has been used whose maximum spindle speed is 1800 rpm, tool feed of 10 m/min, and a 25 KW drive motor.[6]

FIGURE 6.3   CNC milling machine.

6.2.2   EXPERIMENTAL LAYOUT

The experiment was done in a CNC vertical milling machine. Two cutting tools were used tungsten carbide and ceramic tool for machining Inconel 718 which is the workpiece. The experimental run was carried out using L9 orthogonal array for the cutting parameters-spindle speed, tool feed, and depth of cut to determine the response parameters: Surface roughness and tool wear.[4] Table 6.1 shows the input values taken for the experiment run.

TABLE 6.1   Control Factors and Their Levels for the Experiment.

Parameter	Unit	Symbol	Level 1	Level 2	Level 3
Spindle speed	rpm	v	500	600	700
Tool feed	mm/min	f	10	15	20
Depth of cut	mm	d	0.2	0.4	0.6

Tables 6.2 and 6.3 show the observations seen after the experiment was conducted. First, using tungsten carbide as the cutting tool on machining Inconel 718 and secondly using ceramic tool as the cutting tool on Inconel 718. After that Gray relational analysis was used to analyze the input values and find out which experimental run gives low surface roughness and minimum tool wear. Signal to noise ratio (S/N) is another technique used to determine which cutting parameter holds good for determining better response parameters.[5]

6.3 RESULTS AND DISCUSSION

6.3.1 GRAY RELATIONAL ANALYSIS

It is an effective technique by which the relational degree of every factor can be optimized. This method is used to optimize the control parameters and determine the best possible outcome for surface roughness and tool wear.[7] The gray relational grade is found out by following the five steps[6]:

Step-1 Selecting the better value for each influence factor.
Step-2 Obtaining the absolute difference between compared series and referential series.
Step-3 Finding out the minimum and maximum of influence factor
Step-4 Choosing the constant *p*.
Step-5 Calculating the relational co-efficient and relational degree using formulae.

After following these steps, the results obtained for tungsten carbide and ceramics were seen.

From Table 6.2, it can be observed that Experimental run 5 gives a better performance than the rest runs. It is determined using gray relational analysis technique and finding the ranking. Run 5 has a spindle speed of 600 rpm, tool feed of 15 mm/min, and depth of cut 0.6 mm.

From Table 6.3, it can be observed that Experimental run 2 gives a better performance than the rest experimental runs. This was found out by using gray relational analysis technique and finding the ranking of the run. Hence, run 2 being ranked 1 which has a spindle speed of 500 rpm, tool feed of 15 mm/min, and depth of cut 0.4 mm proves to be better than the rest runs.

TABLE 6.2 Relational Degree and Ranking for Tungsten Carbide Tool Insert.

Experimental run	Surface roughness (µm)	Tool wear (mm)	Relational degree	Rank
ω	0.4	0.6	–	–
τ_1	1.538	0.973	1.199	2
τ_2	0.862	0.885	0.876	4
τ_3	0.576	1.000	0.830	5
τ_4	1.000	0.936	0.962	3
τ_5	1.538	1.592	1.570	1
τ_6	0.502	0.973	0.785	6
τ_7	0.707	0.597	0.641	9
τ_8	0.439	0.973	0.759	8
τ_9	0.622	0.885	0.779	7

TABLE 6.3 Relational Degree and Ranking for Ceramics Tool Insert.

Experimental run	Surface roughness (µm)	Tool wear (mm)	Relational degree	Rank
ω	0.4	0.6	–	–
τ_1	0.886	0.834	0.854	2
τ_2	1.000	1.027	1.016	1
τ_3	0.396	1.000	0.758	3
τ_4	1.074	0.385	0.660	5
τ_5	0.412	0.985	0.756	4
τ_6	0.327	0.388	0.364	9
τ_7	0.920	0.418	0.619	6
τ_8	0.307	0.747	0.571	7
τ_9	0.542	0.389	0.450	8

6.3.2 SIGNAL TO NOISE RATIO (S/N)

It is one of the methods to compare the parameters-spindle speed, tool feed, and depth of cut. It even determines the response parameters surface roughness and tool wear to the maximum extent of better results.[8]

From Table 6.4, it can be observed that tool feed has a better feature in determining the surface roughness when tungsten carbide tool insert is being used in machining Inconel 718.

TABLE 6.4 Response Table for S/N Ratios of Surface Roughness Using Tungsten Carbide Tool Insert.

Level	Spindle speed (rpm)	Tool Feed (mm/min)	Depth of Cut (mm)
1	27.65	20.61	26.02
2	23.52	24.89	30.68
3	21.97	27.65	16.44
Delta	5.68	7.04	14.24
Rank	3	2	1

From Table 6.5, it can be observed that the depth of cut has a better feature in determining the tool wear when tungsten carbide tool insert is being used in machining Inconel 718.

TABLE 6.5 Response Table for S/N Ratios of Tool Wear Using Tungsten Carbide Tool Insert.

Level	Spindle speed (rpm)	Tool feed (mm/min)	Depth of cut (mm)
1	4.480	2.841	7.468
2	4.015	5.894	3.847
3	7.700	7.460	4.880
Delta	3.685	4.619	3.622
Rank	2	1	3

From Table 6.6, it can be observed that tool feed has a better feature in determining the surface roughness when ceramics tool insert is being used in machining Inconel 718.

TABLE 6.6 Response Table for S/N Ratios of Surface Roughness Using Ceramics Tool Insert.

Level	Spindle speed (rpm)	Tool feed (mm/min)	Depth of cut (mm)
1	3.2928	4.5556	0.9852
2	1.9058	1.6527	3.9489
3	2.2353	1.2256	2.4998
Delta	1.3871	3.3301	2.9638
Rank	3	1	2

From Table 6.7, it can be observed that spindle speed has a better feature in determining the tool wear when ceramics tool insert is being used in machining Inconel 718.

TABLE 6.7 Response Table for S/N Ratios of Tool Wear Using Ceramics Tool Insert.

Level	Spindle speed (rpm)	Tool feed (mm/min)	Depth of cut (mm)
1	0.2117	16.7116	10.4540
2	20.6348	0.2043	20.0330
3	15.1358	18.6430	5.0720
Delta	20.8465	18.4387	14.9610
Rank	1	2	3

6.4 CONCLUSIONS

Through Gray relational analysis, it was observed that a high spindle speed of 600 rpm can be used for tungsten carbide insert while spindle speed of 500 rpm can be used for ceramics tool insert. Anova predicted the performance of feed rate to play a major role in finding the surface roughness using both inserts-tungsten carbide and ceramics. However, in the case of tool wear, depth of cut is found to perform well for tungsten carbide insert while spindle speed proved a distinguishing parameter for ceramics tool insert. Gray relational analysis and ANOVA showed better tools to determine the control factors and response parameters in determining the machinability of Inconel 718.

ACKNOWLEDGMENTS

The authors deeply appreciate the Centre for Research Projects, CHRIST (Deemed to be University) for providing all the financial aids in executing this research work against the Monograph (MNGDFE-1710). The authors highly solicit the grant received for performing this work and are indebted to the University for providing them lab facilities and research space in doing this work.

KEYWORDS

- **Inconel 718**
- **tungsten carbide**
- **ceramics**
- **S/N**
- **gray relational analysis**

REFERENCES

1. Schornik, V.; Zetek, M.; Dana, M. The Influence of Working Environment and Cutting Conditions on Milling nickel-based Superalloys with Carbide Tools. *Proc. Eng.* **2015,** *100,* 1262–1269.
2. Ma, J.; Wang, F.; Jia, Z.; Xu, Q.; Yang, Y. Study of Machining Parameter Optimization in High Speed Milling of Inconel 718 Curved Surface Based on Cutting Force. *Int. J. Adv. Manu. Technol.* **2014,** *75,* 269--277.
3. Rout, I. S.; Pandian, P. P. A Review on Parametric Study of Hard Machining of Inconel 718 Using Coated Carbide Tool. *Adv. Sci., Eng. and Med.* **2018,** *10,* 234–239.
4. Pandian, P. P.; Raja, P.; Sakthimurugan. Optimization of Cutting Parameters of Thin Ribs in High Speed Machining. *Int. J. Eng. Invent.* (IJEI) **2013,** *2* (4), 62–68.
5. Xavior, A.; Jeyapandiarajan, M.; Madhukar, P. M. Tool Wear Assessment During Machining of Inconel 718. 13th Global Congress on Manufacturing and Management, 2017, 174, pp. 1000–1008.
6. Palpandian, P. P.; Rout, I. S. Parametric Investigation of Machining Parameters in Determining the Machinability of Inconel 718 Using Taguchi Technique and Grey Relational Analysis. *Proc. Comp. Sci.* **2018,** *133,* 786–792, 2018.
7. Ramanujam, R.; Venkatesan, K.; Saxena, V.; Joseph, P. Modeling and Optimization of Cutting Parameters in Dry Turning of Inconel 718 Using Coated Carbide Inserts. International Conference on Advances in Manufacturing and Materials Engineering, Elsevier, 2014, pp. 2550–2559.
8. Doddapattar, N. B.; Puneeth, S. Optimization of Cutting Parameters Using Signal to Noise Ratio for Turning Aluminium Alloy AL7050. *Int. J. Ign. Minds* **2014,** *1* (9), 1–6.

CHAPTER 7

Testing of Surface of Nanolayer Powder Carbide Tungsten After Being Treated by Laser

A. V. SHUSHKOV[1,2], A. Y. FEDOTOV[1,2], and A. V. VAKHRUSHEV[1,2*]

[1]*Department "Mechanics of Nanostructures", Institute of Mechanics, Udmurt Federal Research Center, Ural Division, Russian Academy of Sciences, Izhevsk, Russia*

[2]*Department "Nanotechnology and Microsystems", Technic Kalashnikov Izhevsk State Technical University, Izhevsk, Russia*

Corresponding author. E-mail: vakhrushev-a@yandex.ru

ABSTRACT

The study is devoted to experimental investigations of the elastic modulus and hardness dependences of tungsten carbide deposited by laser high-speed sintering on stainless steel substrate and surface topology. The researches of mechanical properties were fulfilled by the indentation method on studying a complex system of physical and mechanical properties of materials Nanotest 600. The surface topology on the noncontact optical profilometer New View 6300 was investigated. Hardness, Young's modulus increases in the passage trajectory of the laser because of the quenching phenomenon.

7.1 INTRODUCTION

Currently, one of the areas in modern engineering products, subjected to intensive contact loading during operation, is the creation of high-strength and wear-resistant coatings on the surface of materials. The study of the topology, physic, and mechanical properties of thin coatings is given great attention in connection with the creation of new coatings with improved characteristics, depending on the technological mode and method of deposition.[1–8]

One of the promising methods of applying high-strength and wear-resistant powder coatings is the method of laser high-speed sintering of powder coating.[9] This method makes it possible to form high-performance coatings from a powder material deposited on a metal surface by treating the surface with a high-energy laser beam. The disadvantage of this technology is the high porosity of the coating and high surface roughness, especially with the wrong technological mode of surface treatment with a laser. Therefore, it is necessary to choose the optimal technological mode, which would ensure that the roughness is uniformly distributed over the entire surface of the coating and the minimum variation of the mechanical characteristics of the coating material. This is especially important when creating thin films and nanocoatings. This task is difficult since laser sintering of the powder coating material is accompanied by high heating and cooling rates of the material, which leads to the appearance of defects and surface deformation. Therefore, studies of the surface topology of coatings and the mechanical properties of coatings along the path of the laser beam and at the boundary of the area of the coating exposed to the laser beam are highly relevant.

One of the main ways to experimentally study the mechanical properties of thin films, micro- and nanocomposite materials is the nanoindentation method.[10–12] The method of nanoindentation determines the hardness, Young's modulus, and other characteristics of both super hard and soft materials using small loads.[13-17]

The purpose of this work was an experimental study of the mechanical characteristics and roughness of a sample with a powder coating based on tungsten carbide deposited on a stainless-steel substrate, treated with a laser beam.

7.2   EXPERIMENTAL TECHNIQUE

The sample under study was obtained by the method of laser high-speed sintering, which is based on the known technology of selective-laser sintering (SLS) of powders when a mixture of materials with different melting points is subjected to heat treatment. As a result, the material is synthesized with a nanostructure, where the particles are connected by means of a matrix on an organic basis, which quickly evaporates. The modes of laser radiation are selected in such a way as to ensure the melting of only low-melting components. This method creates a coating of any materials. The investigated coating is based on the sintering of ultrafine tungsten carbide powders obtained using the process of mechanical activation. Mechanical activation is the process of grinding in high-energy ball mills. Tungsten carbide powder precipitated on a stainless-steel substrate.

Three-dimensional patterns of the surface of a sample with tungsten carbide powder were determined using a New View 6300 contactless optical profilometer. In order to obtain an image of the sample under study and measure its surface topology, the New View 6300 system scans the sample in a noncontact manner based on white light interference. The light inside the interferometric lens is divided into two beams. The first beam is reflected from the sample under study, and the second is reflected from the high-quality control surface located inside the lens.

Mechanical characteristics are determined on the basis of tests using an indentation method on the complex measuring system Nanotest 600 according to the Oliver–Pharr method.[18] This technique consists in selecting the parameters of a power function describing the experimental dependence of the depth of the indenter and the contact area on the applied force. The calculation of hardness and modulus of elasticity is carried out according to the specified data.

The tests were carried out using the Berkovich indenter, which is a three-sided diamond pyramid with an angle at the apex of $65.3°$ and a radius of curvature of about 200 nm.

The measurement of the mechanical characteristics of the surface layers of tungsten carbide was carried out under a load of 250 mN while the loading rate was 12.5 μm/s. Thus, the time of loading and unloading the point of indentation was 20 s each, the holding time at maximum loading was 10 s. The distance between the indentation points was set to 40 μm. After each implantation of the indenter into the sample during the transition to the next point of indentation, the indenter was retracted from the surface to a distance of 30 μm in order to avoid contact with the surface.

A series of experiments were set up in such a way that it was possible to determine the hardness, the reduced modulus of elasticity, and other mechanical characteristics of the sample under investigation along the laser irradiation trajectory and in the zone between this trajectory and the boundary of the laser passes.

7.3   RESULTS OF EXPERIMENTS

Figure 7.1 shows a photograph of the surface of a powder-coated sample obtained on a New View microscope. The photo clearly shows the grooves corresponding to the path of the laser beam.

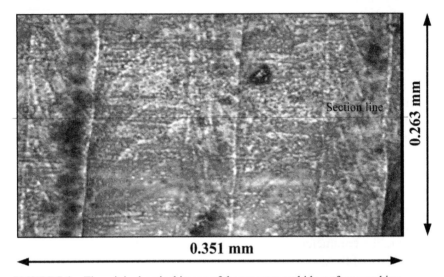

FIGURE 7.1   The original optical image of the tungsten carbide surface coaching.

A three-dimensional picture of the change in the surface profile of tungsten carbide was obtained for this surface on the New View 6300 contactless optical profilometer (Fig. 7.2). The profile of the surface along the section line on Figure 7.1 is shown in Figure 7.3.

The three-dimensional diagram more clearly than the photograph in Figure 7.1 shows the defects in laser processing (surface melting), as well as the trajectory and boundary of the laser passage. High rates of heating and further cooling lead to the fact that only an insignificant part of the powder particles is melted, which provides a mechanically strong sinter layer.

However, a significant part of the powder particles (up to 95%) does not melt, which makes it possible to obtain a sinter layer with preservation of the initial nanostructured, metastable state of the mechanically activated powder.

The tungsten carbide roughness parameter was determined at a section of 0.351×2.63 mm Ra = 1.286 μm.

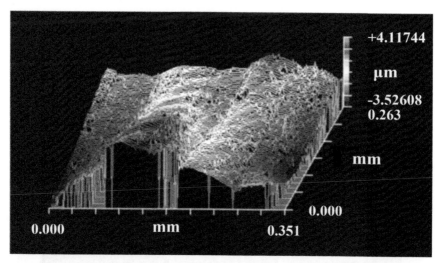

FIGURE 7.2 Three-dimensional picture of the tungsten carbide surface coaching from Figure 7.1.

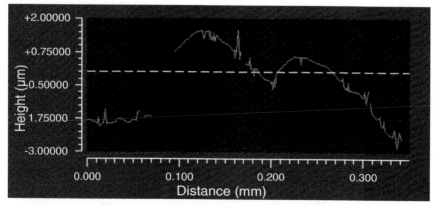

FIGURE 7.3 The profile of the surface along section line on Figure 7.1.

Figure 7.4 shows a photograph of a sample of tungsten carbide and an indenter.

Photographs of the Berkovich indenter prints for indentation points with a loading force of 250 mN are shown in Figure 7.5, and with a double scale in the surface of the sample in Figure 7.6. The photographs clearly show the traces left by the laser and the points of indentation.

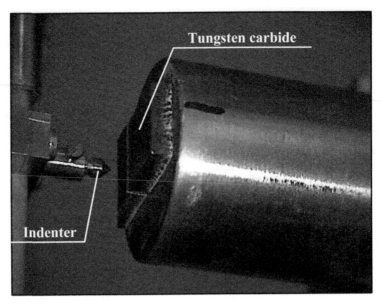

FIGURE 7.4   Photograph of a sample of tungsten carbide and an indenter.

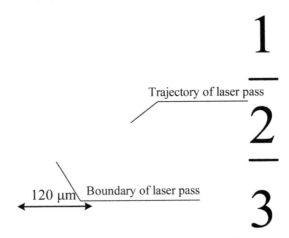

FIGURE 7.5   Photograph of the tungsten carbide surface coaching after indentation: 1, 2, 3, sample coating surfaces treated in various laser beam passes.

Figures 7.7 and 7.8 show that the depth of penetration h_{max} = 1174 nm of the indenter along the path of the laser passage h_{max} = 1321 nm is less than the depth of penetration of the indenter between the path and the boundary of the laser passage.

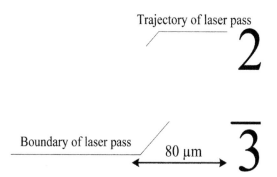

FIGURE 7.6 Photograph of the tungsten carbide surface coaching after indentation: 2, 3, sample coating surfaces treated in various laser beam passes.

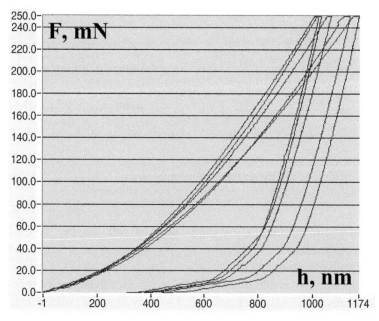

FIGURE 7.7 Dependences of the applied force F, mN from the indenter penetration depth h, nm into the tungsten carbide sample (load-unload diagrams) in laser trajectory; 5 points of indentation.

Thus, the hardness along the laser trajectory should be higher than the hardness in the region between the trajectory and the boundary, as shown in Figure 7.9. We observe a similar dependence for the reduced elastic modulus (Fig. 7.10).

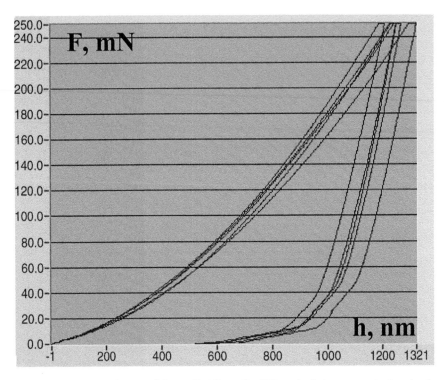

FIGURE 7.8 Dependences of the applied force *F, mN* from the indenter penetration depth *h, nm* into the tungsten carbide sample (load-unload diagrams) in the area between the laser trajectory and the boundary of the laser pass; 5 points of indentation.

According to the graphs (Figs. 7.9 and 7.10) one can see that the hardness and reduced modulus of elasticity of tungsten carbide material along the laser path (phase transformations zone) are higher than the hardness and reduced modulus of elasticity of material in the zone between the trajectory and the boundary of the laser passage. The maximum increase in hardness is 50% and the maximum increase in elastic modulus is 26%, respectively. The averages of the above values differ by 34% and 11%, respectively.

7.4 CONCLUSIONS

The investigations of the effect of laser radiation on the topology and mechanical characteristics of the surface of a coating of powdered tungsten carbide deposited on a stainless-steel substrate have showed:

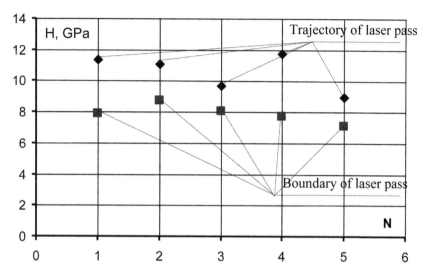

FIGURE 7.9 Experimental values of the hardness *H, GPa*; *N* number of the experiment; 1, hardness in the trajectory of the laser pass; 2, hardness in the area between the trajectory and the boundary of the laser pass.

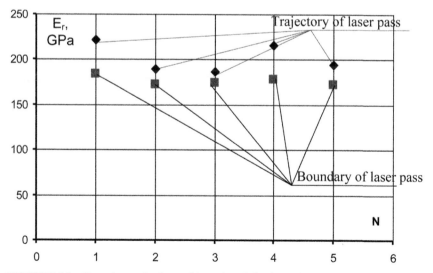

FIGURE 7.10 Experimental values of the reduced elastic modulus E_r, *GPa*; *N* experiment number; 1, elastic modulus on laser trajectory; 2, elastic modulus in the area between the trajectory and the boundary of the laser passes.

1. The structure of the surface is divided into regions and is determined by the trajectory of the laser beam.
2. Surface hardness and modulus of elasticity increase in comparison with the untreated surface by an average of 34% and 11%, respectively.
3. Maximum hardness and strength are observed on the line of the laser beam.

The results of the study can be the basis for selecting the parameters of laser radiation and for optimizing the technology of laser processing of powder refractory coatings.

ACKNOWLEDGMENTS

The authors are grateful to Evgeniy V. Kharanzhevsky, the head of the General Physics Department at Udmurt State University, for their help in preparing the samples.

We are also grateful to Master N. Guryanova for participating in the measurement.

The works was carried out with financial support from the Research Program of the Ural Branch of the Russian Academy of Sciences (project 18-10-1-29) and budget financing on the topic "Experimental studies and multi-level mathematical modeling using the methods of quantum chemistry, molecular dynamics, mesodynamics and continuum mechanics of the processes of formation of surface nanostructured elements and metamaterials based on them" (project 0427-2019-0029).

KEYWORDS

- **powder coating**
- **laser sintering**
- **indentation**
- **Young's modulus**
- **hardness**

REFERENCES

1. Vakhrushev, A. V.; Fedotov, A. Y.; Vakhrushev, A. A.; Shushkov, A. A.; Shushkov, A. V. Investigation of the Metal Nanoparticles Formation Mechanisms, Determination of Mechanical and Structural Characteristics of Nanoobjects and Composite Materials Based on Them. *Chem. Phys. Mesos.* **2010**, *12* (4), 486–495. (In Russian)

2. Shugurov, A. R.; Panin, A. V.; Oskomov, K. V. Features of Determining the Mechanical Characteristics of Thin Films by the Method of Nanoindentation. *Solid State Phys.* **2008**, *50* (6), 1007–1012. (In Russian)

3. Shtansky, D. V.; Kulinich, S. A.; Levashov, E. A.; Moore J. J. Features of the Structure and Physic Mechanical Properties of Nanostructured Thin Films. *Solid State Phys.* **2003**, *45* (6), 1122–1129. (In Russian)

4. Vakhrushev, A. A.; Fedotov A. Y.; Shushkov A.A., Shushkov A.V. Modeling of the Formation of Metal Nanoparticles, the Study of Structural, Physic-mechanical Properties of Nanoparticles and Nanocomposites, *Proceedings of the Tula State University. Natural sciences, a series of "Physics"*, 2011, no. 2, pp. 241–253. (In Russian)

5. Lyakhovich A. M.; Shushkov A. A.; Lyalina N. V. The Strength Properties of Nanoscale Polymer Films Obtained in Benzene Low-temperature Plasma. *Chem. Phys. Mesos.* **2010**, *12* (2), 243–247. (In Russian)

6. Vorobev, V. L.; Bykov, P. V.; Bayankin, V. Y.; Shushkov, A. A.; Vakhrushev, A. V.; Orlova, N. A. Change in Mechanical Properties of Carbon Steel Art. 3 Depending on the Average Current Density in the Beam Under Pulsed Irradiation with Argon Ions, *Phys. Chem. Mat. Proc.* **2012**, *6*, 5–9.

7. Shushkov, A. A.; Vorobyev, V. L.; Vakhrushev, A. V.; Bykov, P. V.; Bayankin, V. Y. Mechanical Properties of Carbon Steel Steel 3 Irradiated with Argon Ions with Different Ion Current Densities], Himicheskaya fizika i mezoskopiya. *Chem. Phys. Mesos.* **2012**, *14* (1), 97–105. (In Russian)

8. Krivilev, M. D.; Kharanzhevsky, E. V.; Gordeev, G. A.; Ankudinov, V. E. Control of Large Systems. *Cont. Tech. Syst. Technol. Proc.* **2010**, *31*, 299–322. (In Russian)

9. Andrievskij, R. A.; Kalinnikov, G. V. et al. Nanoindentation and Deformation Characteristics of Nanostructured Boron-nitride Films. *Solid State Phys.* **2000**, *42* (9), 1624–1627. (In Russian)

10. Jung, Y.-G.; Lawn, B. R.; Huang, H. et al. Evaluation of Elastic Modulus and Hardness of Thin films by Nanoindentation. *J. Mat. Res.* **2004**, *19* (10), 3076–3080.

11. Shojaei, O. R.; Karimi, A. Comparison of Mechanical Properties of Tin Thin Films Using Nanoindentation and Bulge Test. *Thin Solid Films* **1998**, *332* (1–2), 202–208.

12. Sklenika, V.; Kucharova, K. et al. Mechanical and Creep Properties of Electrodeposited Nickel and its Particle-reinforced Nanocomposite. *Rev. Adv. Mater. Sci.* **2005**, *10*, 171–175.

13. Gong, J.; Miao, H.; Peng, Z. A New Function for the Description of the Nanoindentation Unloading Data. *Scripta Materialia.* **2003**, *49* (1), 93–97.

14. Vakhrushev, A. V.; Shushkov, A. V.; Shushkov, A. A. Experimental Study of the Young's Elastic Modulus and Hardness of Iron Microparticles by Indentation. *Chem. Phys. Mesos.* **2009**, 11 (2), 258–262. (In Russian)

15. Cho, S. -J.; Lee, K. -R.; Eun, K. Y. Determination of Elastic Modulus and Poisson's Ratio of Diamond-like Carbon Films. *Thin Solid Films* **1999,** *341* (1–2), 207–210.
16. Vaz, A. R.; Salvadori, M. C.; Cattani, M. Young Modulus Measurement of Nanostructured Palladium Thin Films. *Nanotech.* **2003,** *3*, 177–180.
17. Golovin, Y. I. *Nanoindentation and its Possibilities*. Moskva: Mashinostroenie. 2009. (In Russian)
18. Oliver, W.; Pharr, G. An Improved Technique for Detemining Hardness and Elastic Modulus Using Load and Displacement Sensing Indentation Experiments. *J. Mater. Res.* **1992,** *7* (6), 1564–1583.

CHAPTER 8

Theoretical Analysis of Au–V Nanoalloy Clusters: A Density Functional Approach

PRABHAT RANJAN[1] and TANMOY CHAKRABORTY[2*]

[1]Department of Mechatronics Engineering, Manipal University Jaipur, Jaipur-303007, India

[2]Department of Chemistry, Manipal University Jaipur, Jaipur-303007, India

*Corresponding author. E-mail: tanmoychem@gmail.com; tanmoy.chakraborty@jaipur.manipal.edu

ABSTRACT

A systematic first principle calculation on the electronic properties for small nanoalloy clusters of gold–vanadium is performed using local spin density approximation (LSDA) exchange correlation with basis set LANL2DZ. Density functional theory (DFT) is one of the most popular techniques of quantum mechanics to study the electronic properties of matter. Conceptual density functional theory (CDFT)-based descriptors have spun to be an important key parameter to analyze and correlate the experimental properties of clusters. In this report, small nanoalloy clusters of gold and vanadium, Au_nV ($n = 1$–7), have been studied invoking CDFT-based descriptors. The CDFT-based descriptors viz., highly occupied molecular orbital (HOMO)–lowest unoccupied molecular orbital (LUMO) energy gap, molecular hardness (η), softness (S), electronegativity (χ), and electrophilicity index (ω), have been computed. The linear regression analysis has been done between HOMO–LUMO energy gap and molecular hardness. The high value of regression coefficient predicts the efficacy of our proposed model.

8.1 INTRODUCTION

In recent decade, nanoengineering has progressed as a new research arena in the field of science and technology. Nanoengineering is becoming one of the integral parts of human's life due to its potential applications in the domain of photovoltaics cells, clean-energy industries, photocatalysts, cosmetics, petrochemicals, pharmaceuticals, etc.[1–10] Nanoparticles have been categorized as per their size range, which varies between 1 and 100 nm.[1–5] Due to large surface-to-volume ratio and quantum mechanics effects, nanoparticles offer diverse physico-chemical properties. However, it has been observed in the case of nonlinear alteration that physical and chemical properties of nanoparticles also change due to change in size, shape, and composition.[6] There are a number of scientific reports available in which change in electronic, optical, and magnetic properties is reported due to change in their size, shape, and composition.[1,3,4] The study on nanoparticles is of considerable interest due to its potential technological applications in the field of nanoscience, biophysics, life sciences, nano-electronics, catalysis, etc. [1,3,7–10] A systematic approach of nanoparticles with well-defined geometry and composition may offer some other alternatives for better performance.[8]

Study of gold nanoalloy clusters has attracted a lot of attention due to its considerable electronic and optical properties. The study reveals that gold clusters display different geometry and physico-chemical properties compared to copper and silver at the macroscopic and microscopic level.[1,2] The tremendous work done by scientist Pyykkö,[11–12] Schwerdt-feger, [13] Daniel and Astruc, [2] and Häkkinen[14] in the field of gold nano-clusters reveals that significant physico-chemical properties of gold are produced by enormous relativistic effects. It is already established that geometries, stabilities, and physico-chemical properties of gold clusters can be tuned with the inclusion of dopant atoms ranging from transition metals to nonmetallic elements and alkali metals.[15–18] In order to enhance the stability and physico-chemical properties of gold clusters, various studies have been performed on doped gold nanoalloy clusters. Hansen et al. [19] have studied decay rate and branching ratios of monomer dimer of gold clusters using photo-dissociation technique. Gruene et al.[20] have studied the physical and chemical properties of neutral gold nanoclusters Au_7, Au_{19}, and Au_{20} in the gaseous state. In this chapter, authors revealed that geometry of Au_{20} is similar to bulk gold and have very high-energy gap more than C_{60}. The neutral gold clusters, Au_n, change their state from planar to non-planar at $n = 13$. However, in the case of cationic Au_n^+ and anionic Au_n^- cluster, transition from 2D to 3D occurs at $n = 7$ and $n = 11$, respectively.[21–23]

In order to develop more understanding about gold nanoalloy clusters, it has been observed that study about electronic and optical properties of bimetallic gold-vanadium nanoalloy clusters is very limited. Zhai et al.[24] have reported the structural and electronic properties of anionic gold clusters doped with vanadium, niobium, and tantalum [MAu_{12}^-, (M = V, Nb, Ta)] using photoelectron technique and density functional theory (DFT) methodology. Li et al.[25] have also studied the electronic and magnetic properties of anionic gold clusters doped with titanium, vanadium, and chromium [MAu_6^- (M= Ti, V, Cr)]. In this molecular system, doped atom is placed in the middle of the gold cluster Au_6 ring, which results in high magnetic moments and planar geometry of these clusters. Stener et al.[26] have also investigated photoabsorption spectra of anionic gold clusters doped with vanadium, niobium, and tantalum [MAu_{12}^- (M = V, Nb, and Ta) using time-dependent density functional theory (TDDFT) method. Nhat et al.[27] have studied the physical and chemical properties of vanadium-doped gold clusters, Au_nV ($n = 1–14$) using DFT methodology. The study reveals that addition of vanadium atom enhances the thermodynamic stability of gold clusters.

DFT is one of the important techniques of quantum mechanics to study the electronic properties of matter.[28] Due to computational friendly behavior, DFT is the broadly accepted technique to study the many-body systems. In the domain of material science research, particularly in super conductivity of metal-based alloys, structural, electronic, magnetic, and optical properties of nanoalloy clusters, quantum fluid dynamics, molecular dynamics, and nuclear physics, DFT has received an enormous importance.[29] DFT covers three major domains viz. theoretical, conceptual, and computational.[30–33] Conceptual density functional theory (CDFT) has been proven as an essential tool to study the chemical reactivity of materials.[34–36] The CDFT is highlighted following Parr's dictum: "Accurate calculation is not synonymous with useful interpretation. To calculate a molecule is not to understand it."[37] We have successfully applied DFT technique to study the structures and physico-chemical properties of various nano-composite materials.[38–53]

In this chapter, we have performed a theoretical analysis of gold-vanadium nanoalloy clusters using CDFT approach. The DFT-based descriptors namely, molecular hardness, softness, electronegativity, and electrophilicity index, along with highly occupied molecular orbital (HOMO)–lowest unoccupied molecular orbital (LUMO) energy gap have been computed. The high value of regression coefficient obtained between DFT-based descriptors and HOMO–LUMO energy gap supports our study.

8.2   COMPUTATIONAL DETAILS

In this chapter, we have made an analysis on the physico-chemical properties of bi-metallic nanoalloy clusters of Au_nV ($n = 1–7$). 3D modeling and structural optimizations of all the clusters have been performed using Gaussian 03[54] within DFT framework. For optimization purpose, local spin density approximation (LSDA) exchange correlation with basis set LANL2DZ[55–57] has been implemented. The basis set LANL2DZ has high accuracy for bimetallic clusters, which is already reported previously in the literature.[6,48–51]

Invoking Koopman's approximation,[34] ionization energy (I), and electron affinity (A) of all the nanoalloys have been computed using the following ansatz:

$$I = - \varepsilon_{HOMO} \tag{8.1}$$

$$A = - \varepsilon_{LUMO} \tag{8.2}$$

Thereafter, using I and A, the CDFT-based descriptors viz., electronegativity (χ), global hardness (η), molecular softness (S), and electrophilicity index (ω) have been computed. The equations used for such calculations are as follows:

$$\chi = -\mu = \frac{I + A}{2} \tag{8.3}$$

Where, μ represents the chemical potential of the system.

$$\eta = \frac{I - A}{2} \tag{8.4}$$

$$S = \frac{1}{2\eta} \tag{8.5}$$

$$\omega = \frac{\mu^2}{2\eta} \qquad (8.6)$$

8.3 RESULTS AND DISCUSSIONS

A theoretical analysis of Au_nV ($n = 1$–7) nanoalloy clusters has been performed by using DFT methodology. The orbital energies in form of HOMO–LUMO energy gap along with computed DFT-based global descriptors namely molecular hardness (η), softness (S), electronegativity (χ), and electrophilicity index (ω) have been presented in Table 8.1. The molecular dipole moment in debye unit is also reported in Table 8.1. The result obtained from Table 8.1 reveals that HOMO–LUMO gaps of Au–Vnanoalloy clusters maintain a direct relationship with their calculated hardness values. The trend is anticipated from experimental point of view as the molecular hardness of cluster increases with an increase in frontier orbital energy gap. The molecular system having large energy gap is more stable and less reactive.[31] The result obtained from Table 8.1 shows that Au–V has maximum HOMO–LUMO energy gap, whereas Au_7V has least energy gap in the molecular system of Au_nV ($n = 1$–7). Due to lack of quantitative data of optical properties of these molecular systems, we can consider that there must be a direct qualitative correlation between optical properties of gold-vanadium nanoalloy clusters with their computed HOMO–LUMO energy gap. This is because optical properties of nanomaterials depend on the movement of electrons from valence band to conduction band. On that basis, it may be established that optical properties of Au_nV nanoalloy clusters increase with an increase of their hardness values. Similarly, molecular softness shows an inverse relationship to the optical properties of gold-vanadium clusters. The other DFT-based descriptors viz., electronegativity and electrophilicity index, also display similar correlation with the HOMO–LUMO energy gap. The linear correlation between HOMO-LUMO energy gaps along with calculated electrophilicity index is shown in Figure 8.1. The correlation coefficient, $R^2 = 0.842$ is obtained between electrophilicity index and HOMO–LUMO energy gap. The maximum value of correlation coefficient ($R^2 = 1$) is obtained between molecular hardness and HOMO–LUMO energy gap, which also represents that there is a direct relationship between molecular hardness and HOMO–LUMO energy gap.

TABLE 8.1   Computed DFT-based Descriptors of Au$_n$V (n = 1–7) Nanoalloy Clusters.

Species	HOMO–LUMO gap (eV)	Electronegativity (eV)	Hardness (eV)	Softness (eV)	Electrophilicity index (eV)
AuV	1.214	3.331	0.607	0.824	9.140
Au$_2$V	0.844	5.537	0.422	1.186	36.349
Au$_3$V	0.218	5.034	0.109	4.594	116.408
Au$_4$V	0.299	6.027	0.150	3.341	121.362
Au$_5$V	0.397	5.135	0.199	2.517	66.362
Au$_6$V	0.599	5.769	0.299	1.671	55.588
Au$_7$V	0.210	5.609	0.105	4.773	150.177

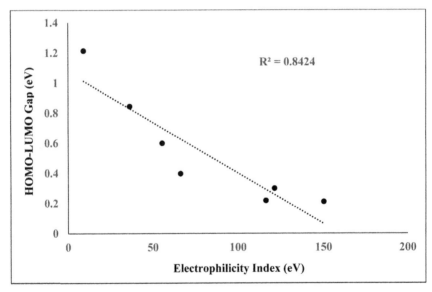

FIGURE 8.1   A linear correlation between electrophilicity index (eV) and HOMO–LUMO gap (eV).

8.4   CONCLUSION

The study of nanoalloy clusters is an important field of research due to its potential applications in renewable energy, nanoelectronics, life sciences, etc. In this chapter, we have reported the physical and chemical properties of gold-vanadium nanoalloy clusters invoking the DFT methodology. The DFT-based descriptors viz., HOMO–LUMO energy gap, molecular hardness, softness, electronegativity, and electrophilicity index, have been computed. The computed data reveal that HOMO–LUMO energy gap of the Au-V clusters has a direct relationship with molecular hardness values and an inverse relation with molecular softness data. Nanoalloy clusters having large HOMO–LUMO energy gap are more stable and less reactive. This is an expected trend from experimental point of view, as hardness of cluster increases with an increase in frontier orbital energy gap. The high value of linear regression coefficient between electrophilicity index and HOMO–LUMO energy gap supports our computational study.

KEYWORDS

- **density functional theory**
- **gold-vanadium**
- **LSDA**
- **HOMO–LUMO energy gap**
- **regression analysis**

REFERENCES

1. Zabet-Khosousi, A.; Dhirani, A.-A. Charge Transport in Nanoparticle Assemblies. *Chem. Rev.* **2008,** *108,* 4072–4124.
2. Daniel, M. C.; Astruc, D. Gold Nanoparticles: Assembly, Supramolecular Chemistry, Quantum-size-related Properties, and Applications toward Biology, Catalysis, and Nanotechnology. *Chem. Rev.* **2004,** *104,* 293–346.
3. Ghosh, S. K.; Pal, T. Interparticle Coupling Effect on the Surface Plasmon Resonance of Gold Nanoparticles: From Theory to Applications. *Chem. Rev.* **2007,** *107,* 4797–4862.
4. Ghosh Chaudhuri, R.; Paria, S. Core/shell Nanoparticles: Classes, Properties, Synthesis Mechanisms, Characterization, and Applications. *Chem. Rev.* **2012,** *112,* 2373–2433.
5. Alivisatos, A. P. Semiconductor Clusters, Nanocrystals, and Quantum Dots. *Sci.-New Series* **1996,** *271,* 933–937.
6. Kastner, M. A. Artificial Atoms. *Phys. Today* **1993,** *46,* 24–31.
7. Gruene, P.; Rayner, D. M., Redlich, B., van der Meer, A. F. G., Lyon, J. T.; Meijer, G.; Fielicke, A. Structures of Neutral Au_7, Au_{19}, and Au_{20} Clusters in the Gas Phase. *Science* **2008,** *321,* 674–676.
8. Haruta, M. Catalysis of Gold Nanoparticles Deposited on Metal Oxides. *Cattech* **2002,** *6,* 102–115.

9. Ismail, R. Theoretical Studies of Free and Supported Nanoalloy Clusters, Ph.D. Thesis. **2012**, 20–38.
10. Roucoux, A.; Schulz, J.; Patin, H. Reduced Transition Metal Colloids: A Novel Family of Reusable Catalysts? *Chem. Rev.* **2002**, *102*, 3757–3778.
11. Pyykkö, P. Theoretical Chemistry of Gold. *Angew. Chem., Int. Ed.*, **2004**, *43*, 4412–4456.
12. Pyykkö, P. Relativistic Effects in Structural Chemistry. *Chem. Rev.* **1988**, *88*, 563–594.
13. Schwerdtfeger, P. Relativistic Effects in Properties of Gold. *Heteroat. Chem.* **2002**, *13*, 578–584.
14. Häkkinen, H. Atomic and Electronic Structure of Gold Clusters: Understanding Flakes, Cages and Superatoms from Simple Concepts. *Chem. Soc. Rev.* **2008**, *37*, 1847–1859.
15. Majumdar, C.; Kandalam, A. K.; Jena, P. Structure and Bonding of Au_5M (M= Na, Mg, Al, Si, P and S) Clusters. *Phys. Rev. B* **2006**, *74*, 205437-1-6.
16. Pyykkö, P.; Runerberg, N. Icosahedral WAu_{12}: A Predicted Closed-shell Species, Stabilized by Aurophilic Attraction and Relativity and in Accord with the 18-Electron Rule. *Angew. Chem., Int. Ed.* **2002**, *41* 2174–2176.
17. Zhai, H. J.; Li, J.; Wang, L. S. Icosahedral Gold Cage Clusters: $M@Au_{12}^-$ (M = V, Nb, and Ta). *J. Chem. Phys.* **2004**, *121*, 8369–8374.
18. Li, X.; Kiran, B.; Li, J.; Zhai, H. J.; Wang, L. S. Experimental Observation and Confirmation of Icosahedral $W@Au_{12}$ Molecules. *Angew. Chem., Int. Ed.* **2002**, *41*, 4786–4789.
19. Hansen, K.; Herlert, A.; Schweikhard, L.; Vogel, M. Dissociation Energies of Gold Clusters Au_N^+, N = 7–27. *Phys. Rev. A* **2006**, *73*, 063202.
20. Gruene, P.; Rayner, D. M.; Redlich, B.; Van der Meer, A. F. G.; Lyon, J. T.; Meijer, G.; Fielicke, A. Structures of Neutral Au_7, Au_{19} and Au_{20} Clusters in the Gas Phase. *Sci.* **2008**, *321*, 674–676.
21. Gilb, S.; Weis, P.; Furche, F.; Ahlrichs, R.; Kappes, M. M. Structures of Small Gold Clusters Cations (Au_n^+, $n<14$): Ion Mobility Measurements versus Density Functional Calculations. *J. Chem. Phys.* **2002**, *116*, 4094–4101.
22. Woldeghebriel, H.; Kshirsagar, A. How Cationic Gold Clusters Respond to a Single Sulfur Atom. *J. Chem. Phys*, **2007**, *127*, 224708.
23. Furche, F.; Ahlrichs, R.; Weis, P.; Jacob, C.; Gilb, S.; Bierweiler, T.; Kappes, M. M. The Structures of Small Gold Cluster Anions as Determined by a Combination of Ion Mobility Measurements and Density Functional Calculations. *J. Chem. Phys.* **2002**, *117*, 6982–6990.
24. Zhai, H. J.; Li, J.; Wang, L. S. Icosahedral Gold Cage Clusters: $M@Au_{12}^-$ (M=V, Nb, and Ta). *J. Chem. Phys.* **2004**, *121*, 8369–8374.
25. Li, X.; Kiran, B.; Cui, L. F.; Wang, L. S. Magnetic Properties in Transition-metal-doped Gold Clusters: $M@Au_6$ (M=Ti, V, Cr). *Phys. Rev. Lett.* **2005**, *95*, 253401.
26. Stener, M.; Nardelli, A.; Fronzoni, G. Theoretical Study on the Photoabsorption of Mau_{12}^- (M= V, Nb, and Ta). *Chem. Phys. Lett.* **2008**, *462*, 358–364.
27. Nhat, P. V.; Nguyen, M. T. Trends in Sturcutral, Electronic and Energetic Properties of Bimetallic Vanadium-Gold Clusters Au_nV with n = 1–14. *Phys. Chem. Chem. Phys.* **2011**, *13*, 16254–16264.
28. Cramer, C. J.; Truhlar, D. G. Density Functional Theory for Transition Metals and Transition Metal Chemistry. *Phys. Chem. Chem. Phys.* **2009**, *11*, 10757–10816.

29. Hafner, J.; Wolverton, C.; Ceder, G. Towards Computational Materials Design: The Impact of Density Functional Theory of Materials Research. *MRS Bull.* **2006**, *31*, 659–668.

30. Parr, R. G.; Yang, W. Density Functional Theory of the Electronic Structure of Molecules. *Annu. Rev. Phy. Chem.* **1995**, *46*, 701–728.

31. Kohn, W.; Becke, A. D.; Parr, R. G. Density Functional Theory of Electronic Structure. *J. Phys. Chem.* **1996**, *100*, 12974–12980.

32. Liu, S.; Parr, R. G. Second-order Density-Functional Description of Molecules and Chemical Changes. *J. Chem. Phys.* **1997**, *106*, 5578–5586.

33. Ziegler, T. Approximate Density Functional Theory as a Practical Tool in Molecular Energetics and Dynamics. *Chem. Rev.* **1991**, *91*, 651–667.

34. Parr, R. G.; Yang, W. *Density Functional Theory of Atoms and Molecules*. Oxford: Oxford University Press, 1989.

35. Chermette, H. Chemical Reactivity Indexes in Density Functional Theory. *J. Comput. Chem.* **1999**, *20*, 129–154.

36. Geerlings, P., Proft, F. D., Langenaeker, W. Conceptual Density Functional Theory. *Chem. Rev.*, Washington D.C. **2003**, *103*, 1793–1874.

37. Geerlings, P., Proft, F. D. Chemical Reactivity as Described by Quantum Chemical Methods. *Int. J. Mol. Sci.* **2002**, *3*, 276–309.

38. Ranjan, P.; Dhail, S.; Venigalla, S.; Kumar, A.; Ledwani, L.; Chakraborty, T. A Theoretical Analysis of Bi-metallic (Cu-Ag) n = 1–7 Nano Alloy Clusters Invoking DFT Based Descriptors. *Mater. Sci.-Pol.* **2015**, 33, 719–724.

39. Ranjan, P.; Venigalla, S.; Kumar, A.; Chakraborty, T. Theoretical Study of Bi-metallic Ag_mAu_n (m+n=2–8) Nano Alloy Clusters in Terms of DFT Based Descriptors. *New Front. Chem.* **2014**, *23*, 111–122.

40. Venigalla, S.; Dhail, S.; Ranjan, P.; Jain, S.; Chakraborty, T. Computational Study about Cytotoxicity of Metal Oxide Nanoparticles Invoking Nano-QSAR Technique. *New Front. Chem.* **2014**, *23*, 123–130.

41. Ranjan, P.; Kumar, A.; Chakraborty, T. Computational Study of $AuSi_n$ (n = 1–9) Nanoalloy Clusters Invoking DFT Based Descriptors. *AIP Conf. Proc.* **2016**, *1724*, 020072.

42. Ranjan, P.; Kumar, A.; Chakraborty, T. Theoretical analysis: Electronic and Optical Properties of Gold-Silicon Nanoalloy Clusters. *Mat. Today Proc.* **2016**, *3*, 1563–1568.

43. Ranjan, P.; Kumar, A.; Chakraborty, T. Computational Study of Nanomaterials Invoking DFT Based Descriptors. In *Environmental Sustainability: Concepts, Principles, Evidences and Innovations*; Mishra, G. C., Ed.; , Excellent Publishing House: New Delhi, 2014; pp 239–242.

44. Ranjan, P.; Venigalla, S.; Kumar, A.; Chakraborty, T. A Theoretical Study of Bi-metallic AgAun (n = 1–7) Nano Alloy Clusters Invoking DFT Based Descriptors. In *Recent Methodology in Chemical Sciences: Experimental and Theoretical Approaches*; Chakraborty, T.; Ledwani, L. Eds.; Apple Academic Press and CRC Press: USA, 2015; pp 337–346.

45. Ranjan, P.; Kumar, A.; Chakraborty, T. Computational Investigation of Ge Doped Au Nanoalloy Clusters: A DFT Study. *IOP Conf. Series: Mater. Sci. Eng.* **2016**, *149*, 012172.

46. Dhail, S.; Ranjan, P.; Chakraborty, T. Correlation of the Experimental and Theoretical Study of Some Novel-2-phenazinamine Derivatives in Terms of DFT Based Descriptors. In *Crystallizing Ideas- The Role of Chemistry*; Ramasami, P.; Bhowon, M. G.; Laulloo, S. J.; Wah H. L. K., Eds.; Springer: Switzerland, 2016; pp 97–112.

47. Ranjan, P.; Kumar, A.; Chakraborty, T. Theoretical Analysis: Electronic and Optical Properties of Small Cu-Ag Nano Alloy Clusters. In *Computational Chemistry Methodology in Structural Biology and Material Sciences*; Chakraborty, T.; Ranjan, P.; Pandey, A. Eds.; Apple Academic Press and CRC Press: USA (In Press) ISBN- 9781315207544.

48. Ranjan, P.; Chakraborty, T.; Kumar, A. A Theoretical Study of Bimetallic CuAuN (N = 1–7) Nanoalloy Clusters Invoking Conceptual DFT-based Descriptors. In *Applied Chemistry and Chemical Engineering*; Haghi, A. K.; Pogilani, L.; Castro, E. A.; Balkose, D.; Mukbaniani, O. V.; Chia C. H., Eds.; Vol. 4, Apple Academic Press and CRC Press, USA (In Press), ISBN- 9781315207636.

49. Ranjan, P.; Chakraborty, T.; Kumar, A. Computational Investigation of Cationic, Anionic and Neutral Ag_2Au_N (N = 1–7) Nanoalloy Clusters. *Phy. Sci. Rev.*, **2007**, 1–13.

50. Ranjan, P.; Chakraborty, T. A DFT Study of Vanadium Doped Gold Nanoalloy Clusters. *Key Eng. Mater.* **2018**, *777*, 183—189.

51. Ranjan, P.; Chakraborty, T.; Kumar, A. Computational Study of Au Doped Cu Nano Alloy Clusters. *Nano Hybrids.* **2017**, *17*, 62—71.

52. Ranjan, P.; Chakraborty, T. Density Functional Approach: To Study Copper Sulfide Nanoalloy Clusters. *Acta Chim. Slov.* **2019**, *66*, 173–181.

53. Ranjan, P.; Chakraborty, T.; Kumar, A. Theoretical Analysis: Electronic, Raman, Vibrational and Magnetic Properties of Cu_nAg (n = 1–12) Nanoalloy Clusters. In *Theoretical and Quantum Chemistry at the Dawn of the 21st Century*; Chakraborty, T.; Carbo-Dorca, R. (Eds.; Apple Academic Press: New York, USA, 2018; pp 1–34, ISBN 9781351170956.

54. Gaussian 03, Revision C.02, M. J. Frisch et al., Gaussian Inc., Wallingford, CT, **2004**.

55. Jones, R. O.; Gunnarsson, O. The Density Functional Formalism, Its Applications and Prospects. *Rev. Mod. Phys.* **1989**, *61*, 689—746.

56. Zupan, A.; Blaha, P.; Schwarz, K.; Perdew, J. P. Pressure-induced Phase Transitions in Solid Si, SiO_2, and Fe: Performance of Local-Spin-Density and Generalized-Gradient-Approximation Density Functionals. *Phys. Rev. B* **1998**, *58*, 11266—11272.

57. Stadler, R.; Gillan, M. J. First- Principles Molecular Dynamics Studies of Liquid Tellurium. *J. Phys.: Condens. Matter* **2000**, *12*, 6053–6061.

CHAPTER 9

Analysis of the Uniformity of Mixing of Micro- and Nanoelements

A. V. VAKHRUSHEV[1,2*] and A. V. ZEMSKOV[2]

[1]Department "Mechanics of Nanostructures", Institute of Mechanics,
Udmurt Federal Research Center, Ural Division,
Russian Academy of Sciences, Izhevsk, Russia

[2]Department "Nanotechnology and Microsystems",
Technic Kalashnikov Izhevsk State Technical University, Izhevsk, Russia

*Corresponding author. E-mail: Vakhrushev-a@yandex.ru

ABSTRACT

The description of software–hardware complex for the analysis of the uniformity of mixing of micro- and nanoelements based on a mathematical model to determine the uniformity of mixing of two-component mixture is presented. We formulate the algorithm for calculating the uniformity of mixing two-component mixture of micro- and nanoelements. Results of experimental studies and theoretical analysis of mixing processes of nanotubes and nanodiamonds in time are described.

9.1 INTRODUCTION

One of the significant technological achievements of modern science and technology was the creation of nanocomposite materials containing nanoscale elements of various types: Nanoparticles, nanotubes, nanofibers, etc., as well as microelements.[1–5] At the moment, nanoelements are becoming more widely used in the production of advanced technology of nanomaterials and nanocoating. The effect of the introduction of nanoelements is most clearly observed in the development of multifunctional materials. Using different types of nanoelements and varying their shape, size, density, and the method of their introduction into the final product or semifinished product, you can get materials with a variety of properties. Nanocomposites are used as high-strength materials, used in many fields of technology. Such materials can be 10 times stronger than steel, and have a low mass, have high hardness, high fire resistance, and electrical conductivity.[8]

The creation of such composites is associated with a number of technological problems, one of which is the mixing of nanoelements. The process of mixing nanoelements substantially depends both on the environment in which it takes place and on the phenomena of self-organization of nanoelements. With the wrong choice of mixing medium, agglomeration of nanoparticles occurs, and a homogeneous mixture is not formed. Self-organization, on the one hand, provides effective mixing at the nanoscale level; however, in a number of cases, the occurrence of ordered stable multilevel structures hinders the mixing process.[9–10]

It should be noted that the uneven distribution of nanoelements and a significant change in their average size can, in contrast to conventional composite materials, cause local deterioration of the properties and parameters of the nanocomposite. This is explained by the fact that their properties strongly depend on their size[11–12] and, accordingly, macroscopic properties substantially depend on the uniform distribution of nanoelements in the bulk of the material. Therefore, even a slightly uneven distribution of the components of nanocomposites is unacceptable, and the task of studying the formation of homogeneous nanodisperse mixtures and the operational control of the parameters of the homogeneity of their mixing is very relevant.[13–14]

The aim of this work is to develop an algorithm for calculating the uniformity of mixing of a two-component mixture of micro- and nanoelements and the creation of a software and hardware complex that allows for operative control of the mixing process.

9.2   MATHEMATICAL MODEL OF MIXING PROCESS

The parameter by which one can judge the quality of the mixing of the mixture is the degree of mixing I. The degree of mixing, in general, should be understood as the mutual distribution of two or more substances after mixing the entire system. Various formulas are used to calculate the degree of mixing. The most commonly used formula is Hixon and Tenney[15]:

$$I = \frac{X_1 + X_2 + ... + X_n}{n}$$

Here n is the number of samples taken; X_1, X_2 are relative concentrations of samples taken, calculated by the formulas:

$$X_i = \frac{\Phi_i}{\Phi_{i0}} \quad (\Phi_i < \Phi_{i0})$$

or

$$X_i = \frac{1 - \Phi_i}{1 - \Phi_{i0}} \quad (\Phi_i > \Phi_{i0}),$$

where Φ_i, Φ_{i0} are the volume fractions of the analyzed component in the *i*-th sample and in the entire mixing apparatus, respectively.

In addition, there are many statistical methods for assessing the degree of mixing of mixtures based on the analysis of samples taken.[19]

This paper is devoted to the development of a new method for determining the uniformity of the mixing of a two-component mixture. The difference of this model from others used to determine the uniformity of mixing[1–3] is that it considers the spatial arrangement of particles.

The initial data for the calculation will be a digital snapshot of the mixture with a clear color separation of the components. The picture can be obtained by any method: using an optical microscope, diffraction methods, scanning probe microscopy, etc.[10] The main requirement for the analyzed image is the color difference between the mixed components. The result of the calculation will be a number that quantitatively characterizes the quality of mixing; the degree of mixing *I*.

In order to simplify the calculation algorithm, we introduce some assumptions:

1. Model mixing one-dimensional. The degree of mixing is determined not for the entire sample of the mixture as a whole, but in a single direction. The image of the mixture is cut and along it is calculated mixing. For a snapshot, several such segments are calculated, and then, using the results obtained, the arithmetic average value is calculated, which is the characteristic of the mixing quality of this sample.

2. Particle size is not considered in the calculation, each pixel of the image is taken as a separate particle.
3. The stirred mixture ratio of the components is not considered.

The particles of substances that make up the mixture of different materials are denoted by black and white squares-pixels (Fig. 9.1).

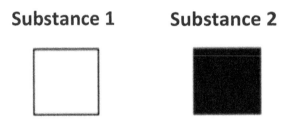

FIGURE 9.1   Symbol of particles of various materials of the composite mixture.

The first step in calculating the degree of mixing of the composite mixture is the formation of the calculated segment. The calculated segment is an array containing pixel data on the location of the mixed particles. To do this, two points $A(x_1,y_1)$ and $B(x_2,y_2)$ are selected in the picture and a segment AB is drawn between them (Fig. 9.2a).

The image pixels that were captured by the drawn segment form a preliminary computational array (Fig. 9.2b) containing data on the location of the mixture particles in the direction of the segment, but not yet suitable for the calculation.

The color of each pixel determines its "belonging" to the background or one of the mixed substances. Background pixels are excluded from the resulting array, since, due to the nature of the computational model, they are not involved in the calculation. Their presence depends only on the method of sampling (for example, when the microscope slides under the microscope, the gaps between the particles increase, but their relative position does not change, and in the case of dense filling or using other methods of imaging, there are no gaps).

As a result, we obtain a calculated array (Fig. 9.2c), according to which the degree of mixing along this segment is determined.

In the case of uniform mixing, the distances between the particles of a substance are equal to each other and have a maximum value (Fig. 9.3).

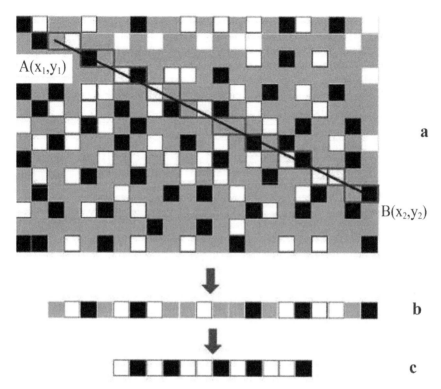

FIGURE 9.2 The scheme of formation of the calculated segment for calculating the uniformity of mixing of the composite mixture: The background color is marked with gray, the mixture particles are marked with black and white, respectively. (a) A diagram of building a segment, (b) a preliminary calculated segment, and (c) a calculated segment.

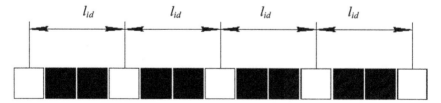

FIGURE 9.3 Distances between particles with their ideal uniform mixing.

The distance between the particles, in this case, is determined by the formula (9.1).

$$l_{id} = \frac{L}{n} - 1 \tag{9.1}$$

where l_{id}, distance between particles with perfect mixing, (pixels);
    L, length of the calculated segment, (pixels);
    n, the number of particles of the mixed powder, (number).

For a real calculated segment, the components are arranged randomly and the distance between the nearest particles (Fig. 9.4) in this area is determined by the formula (9.2).

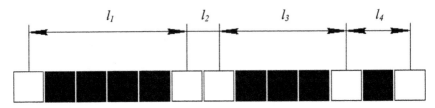

FIGURE 9.4   The lengths of the plots for the real calculation segment.

$$l_i = x_i - x_{i-1} - 1 \tag{9.2}$$

where l_i, distance between particles, (pixels);
    x_{i-1}, x_i, particle coordinates, (pixels).

We introduce a parameter that will characterize the deviation of the distance l_i from the value of l_{id}. Denote it by R_i and call it a deviation from uniformity. If the length of the segment l_i is equal to the distance between particles with ideal mixing l_{id}, the value of R_i is equal to 1. The greater the difference between the lengths of the segments, the smaller the value of R_i.

Given the above, the formula for determining the value of the uniformity of mixing can be represented as a mathematical relationship (9.3).

$$R_i = \begin{cases} \dfrac{l_i}{l_{id}}, & if\ l_i \le l_{id} \\[2ex] \dfrac{l_i - L}{l_{id} - L}, & if\ l_i > l_{id} \end{cases} \tag{9.3}$$

Graphically, the dependence of relations (9.3) is shown in Figure 9.5.

If we find the arithmetic average of all R_i for the considered computational segment, then we get some value I' calculated by formula (9.4). This variable can characterize the uniformity of mixing.

FIGURE 9.5   The graph of the deviation from the uniformity of R depending on the distance between the particles.

$$I' = \frac{1}{N}\sum_{i=1}^{N} R_i \qquad (9.4)$$

where I', the degree of mixing;
         N, number of intervals.

Since one pixel in the model is taken as a separate particle, and each particle in the picture consists of a set of pixels, the result, even in the case of complete mixing, will be less than one, reaching I_{max} instead of 1 (Fig. 9.6). To eliminate this drawback, it is necessary to introduce a correction factor, which will make the necessary amendment.

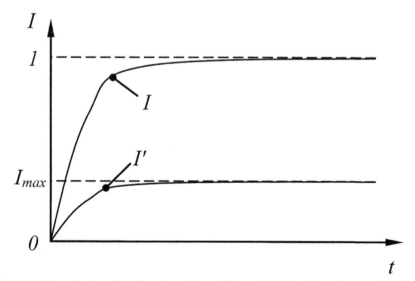

FIGURE 9.6   Graphs of the time dependence of the calculated and true degree of mixing.

To find this coefficient, it is necessary to determine empirically the value of I_{max}, and then calculate the degree of mixing by the formula (9.5).

$$I = \frac{1}{I_{max}} I' \qquad (9.5)$$

9.3 DESCRIPTION OF THE SOFTWARE AND HARDWARE COMPLEX

On the basis of the described algorithm, a program was developed that allows to automate the process of determining the degree of mixing. The interface of the program complex is presented in Figure 9.7.

Consider the work of this software package. In the area of image 1 (Fig. 9.7), the investigated image is loaded. At this stage, you can specify the area of the calculation. After the calculations, this area of the image also displays the location of the calculated segments. Further, using panel 2 (Fig. 9.7), the color filter is adjusted, which allows determining which component of the mixture corresponds to a particular image pixel.

Working with the filter presents some difficulties, since the edges of the particles, due to diffraction and inaccuracies in the adjustment of the microscope, are not clear, but look blurred and brighter than the particles. Panel 8 sets the parameters required for the calculation. Panel 8 also displays the results. The calculation can be carried out randomly (Fig. 9.8a), uniformly over the grid (Fig. 9.8b), or along a straight line in the selection direction defined by the angle α (Fig. 9.8c).

9.4 EXPERIMENTAL TECHNIQUE

Figure 9.9 shows experimental equipment for conducting practical experiments on the mixing of nanopowders and taking pictures of the mixture. The sequence of the stages of the experiment is shown in Figure 9.10.

The study includes five main steps:

1. Prepare the components.
   This stage includes the determination of the required ratio of components, their weighing and the formation of a mixture.

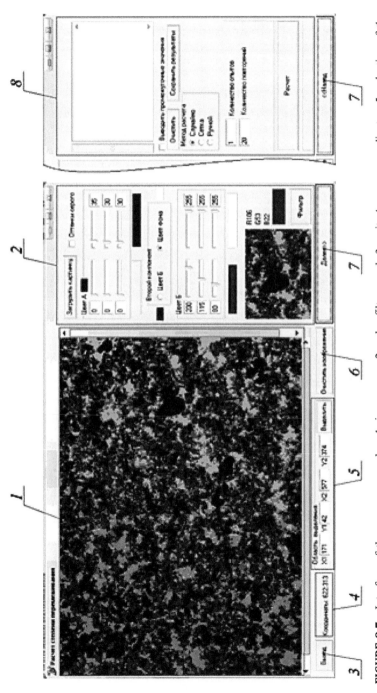

FIGURE 9.7 Interface of the program complex: 1, image area; 2, color filter panel; 3, exit; 4, cursor coordinates; 5, selection of the calculated area; 6, image cleaning; 7, switching between panels; 8, calculation panel.

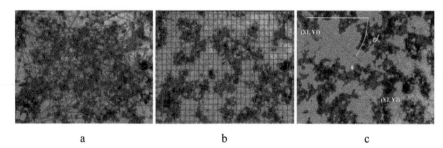

a   b   c

FIGEURE 9.8 Calculation methods: a, randomly; b, uniformly over the grid; c, along a straight line.

FIGURE 9.9 Hardware complex for analyzing the mixing process of micro- and nanoelements: 1, a camera; 2, LCD-TV; 3, optical microscope; 4, ultrasonic bath.

2. Stirring.

   Mixing can be done in any way (mechanical, ultrasonic, etc.). During the experiment, an ultrasonic bath was used.

3. Sampling.
   At certain points in time, samples are taken from the mixture to determine and control the uniformity of mixing.

4. Taking a digital image required for the calculation.
   The picture can be obtained in any way, the main requirement for the image is the color difference of the mixed components.

5. Calculate the degree of mixing.
   At this stage, the resulting images are prepared to work with the program. The image is preprocessed (cropping, color filtering), then the calculation and determination of the degree of mixing is performed.

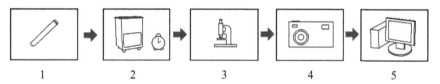

1   2   3   4   5

FIGURE 9.10   The sequence of the experiment.

9.5   RESULTS OF EXPERIMENTS AND THEIR ANALYSIS

The object of the study was the mixing processes of two powders: Nanotubes and nanodiamonds. In total, three series of mixing experiments were performed. In each experiment, three samples were taken, the degree of mixing was calculated, and the average value was determined for each experiment.

Mixing was carried out in an ultrasonic bath, then a sample was taken with a pipette. It was placed on the glass and a digital image was taken from an optical microscope (Fig. 9.11).

Note that the selected magnification of the microscope makes it possible to investigate mixing only groups of different nanoelements since individual elements cannot be identified by an optical microscope. However, the degree of mixing of these elements is determined quite clearly. It can be seen that all the photographs except the first one practically does not differ from each other, and it is impossible to visually determine the mixing uniformity. However, the degree of mixing of the studied mixture varies from snapshot to snapshot, as indicated in Figure 9.12 dependence of the degree of mixing of the mixture from the time of exposure to ultrasound. It is seen that the averaged value of the degree of mixing of the mixture has the form of a smooth, constantly increasing curve (the line without markers in Figure 9.12).

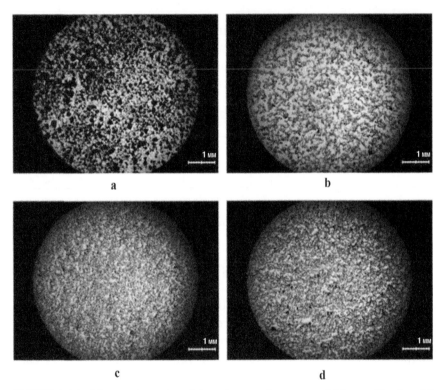

FIGURE 9.11   Photos of samples of the mixture, taken at a tenfold magnification: (a) mixture of powders without mixing; (b–d) mixture of powders after 10, 20, and 30 min of mixing, respectively.

As follows from the graph, the most intensive mixing occurs in the first 15–20 min of the process, then the mixing rate decreases, the degree of mixing stabilizes, and the graph approaches the horizontal asymptote close to 1. A further increase in the degree of mixing by increasing the time of the mixing process is very inefficient. Thus, experiments have shown that the ultrasonic mixing of nanoelements is effective, but it has quite limited time efficiency.

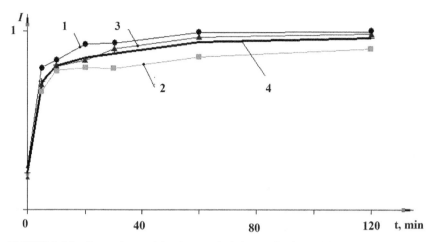

FIGURE 9.12 Dependence of the degree of mixing on the time of the composite mixture of nanodiamonds and nanotubes: 1, 2, 3, various samples of the mixture from different parts of the mixture of elements; 4, the average value of the degree of mixing.

9.6 CONCLUSIONS

The hardware–software complex for analyzing the uniformity of mixing micro- and nanoelements using a mathematical model to determine the uniformity of mixing of a two-component mixture presented in the work can be effectively used in modern nanotechnologies for the operational control of the degree of mixing of micro- and nanocomposite mixtures.

The development of the complex involves building a model of the degree of mixing of three or more micro- and nanocomposite mixtures, using artificial intelligence methods to speed up the process of controlling the mixing of composite mixtures and increasing the degree of automation of this process.

ACKNOWLEDGMENTS

The work was supported by IzSTU (project 28.04.01/18VAV), Research Program of the Ural Branch of the Russian Academy of Sciences (project 18-10-1-29), and UFRC (project 0427-2019-0029).

KEYWORDS

- **uniformity of mixing**
- **mathematical modeling**
- **software–hardware complex**
- **mixing of micro- and nanoelements**

REFERENCES

1. Wei, D.; Dave, R.; Pfeffer, R. Mixing and Characterization of Nanosized Powders: An Assessment of Different Techniques. *J. Nanopart. Res.* **2002,** *4* (1–2), 21–41.
2. Nakamura, H.; Yanagihara, Y.; Sekiguchi, H. et al. Effect of Particle Size on Mixing Degree in Dispensation. *Yakugaku Zasshi* **2004,** *124* (3), 135–139.
3. Nakamura, H.; Yanagihara, Y.; Sekiguchi, H. et al. Effect of Mixing Method on the Mixing Degree During the Preparation of Triturations. *Yakugaku Zasshi* **2004,** *124* (3), 127–134.
4. Vakhrushev, A. V.; Kodolov, V. I.; Haghi, A. K.;Ameta, S. C., Eds.; *Carbon Nanotubes and Nanoparticles. Current and Potential Applications, Series: AAP Research Notes on Nanoscience and Nanotechnology*; Apple Academic Press: Waretown, USA; 2019; p. 291.
5. Vakhrushev, A. V. *Computational Multiscale Modeling of Multiphase Nanosystems. Theory and Applications*; Apple Academic Press: Waretown, New Jersey, USA, 2017; p. 402.
6. Andrievski, R. A.; Ragulia, A. V. *Nanostructural Materials*; Academia: Moscow; 2005; p. 192.
7. Imry, Y. *Introduction to Mesoscopic Physics*; University Press: Oxford; 2002; p. 304.
8. Gusev, A. I.; Rempel, A. A. *Nanocrystalline Materials*; Physical Mathematical Literature: Moscow; 2001; p. 224.
9. Vakhrouchev, A. V. Simulation of Nano-elements Interactions and Self-assembling. *Model. Simul. Mat. Sci. Eng.* **2006,** *14*, 975–991.
10. Dingreville, R.; Qu, J.; Cherkaoui, M. Surface Free Energy and its Effect on the Elastic Behavior of Nano-sized Particles, Wires and Films. *J. Mech. Phys. Solids* **2004,** *53*, (8), 1827–1854.
11. Duan, H. L.; Wang, J.; Huang, Z. P.; Karihaloo, B. L. Size-dependent Effective Elastic Constants of Solids Containing Nano-inhomogeneities with Interface Stress. *J. Mech. Phys. Solids* **2005,** *53* (7), 1574–1596.
12. Ni, Q. Q.; Fu, Y.; Iwamoto, M. Evaluation of Elastic Modulus of Nano Particles in PMMA/Silica Nanocomposites. *J. Soc. Mat. Sci.* **2004,** *53* (9), 956–961.

13. Vakhrushev, A. V.; Fedotov, A. Y.; Vakhrushev, A. A.; Golubchikov, V. B.; Givotkov, A. V. Multilevel Simulation of the Processes of Nanoaerosol Formation. Part 2. Numerical Investigation of the Processes of Nanoaerosol Formation for Suppression of Fires. *Int. J. Nanomech. Sci. Technol.* **2011,** *2* (3), 205–216.

14. Vakhrushev, A. V.; Fedotov, A. Y.; Vakhrushev, A. A.; Suyetin, M. V. The Method and Apparatus of Mixing Nanoparticles//Patent of the Russian Federation 2301771 on 27/06/2007.

15. Hixon, A. W.; Tenney A. H. *Trans. Am. Inst. Chem. Eng.* **1935,** *31,* 113–127.

CHAPTER 10

Simultaneous Multicriteria-Based Optimization Trends in Industrial Cases

ZAINAB AL ANI[1], ASHISH M. GUJARATHI[*],
G. REZA VAKILI-NEZHAAD[1], CHEFI TRIKI[2], and TALAL AL WAHAIBI[1]

[1]Department of Petroleum & Chemical Engineering,
College of Engineering, Sultan Qaboos University, P.O. Box 33,
Al-Khod, P.C. 123, Muscat, Sultanate of Oman

[2]Department of Mechanical and Industrial Engineering,
College of Engineering, Sultan Qaboos University, P.O. Box 33,
Al-Khod, P.C. 123, Muscat, Sultanate of Oman

[*]Corresponding author. E-mail: ashishg@squ.edu.om

ABSTRACT

Sulfuric acid (H_2SO_4) is one of the most essential produced chemicals around the world, as it is used in various applications. The manufacturing process encounters many problems regarding its impact on the environment, like the emissions of sulfur and carbon oxide gases. In the present study, the aim is to minimize the emissions by minimizing the acidification potential (AP) and global warming potential (GWP). For better improvements in the processes' performance, the other two economic objectives are involved simultaneously in two separate optimization cases. The process simulation was done with Aspen Plus with excel-based multiobjective optimization for minimization of AP and total production cost (TPC), and maximization of product sales (PRS) and minimization of GWP. Some operating conditions (decision variables) were selected to carry out the multiobjective optimization (MOO) such as absorbers' and reactors' pressure and raw material flow rates. It was noticed that for the case of minimization of AP and TPC, airflow rate was the most affecting decision variable, whereas the water flow rate decision variable was the key in the case considering the minimization of GWP and maximization of PRS. Results exhibited the potential for MOO for H_2SO_4 production plant for the considered. Flow rates showed the highest effect on the results.

10.1   INTRODUCTION

Sulfuric acid, H_2SO_4, is produced more than any other chemical globally. About 150 million tonnes is the total production of sulfuric acid.[1] It is used in fertilizers' manufacturing and many other applications. The most used raw material in H_2SO_4 production is sulfur.[1] H_2SO_4 production plants cause lots of environmental problems. One of them is the acidic gas emissions like sulfur oxides (SO_x).[2] Kiss et al.[3] and Telang[4] studied the opportunity for optimizing the process from different aspects like maximization of profit, minimization of energy, and minimization of SO_x emissions. Both studies were able to develop the process for the considered objectives.

The acidic emissions are harmful to the environment and living creatures. This impact can be determined by what is called acidification potential (AP).[5,6] Minimizing AP will lead to a more environmental-friendly process. For a more economically efficient plant, another economical objective can be simultaneously optimized with AP, like profit, utility cost, etc.

Due to the impurities in the elemental sulfur utilized, H_2SO_4 production plants are generating carbon dioxide (CO_2), one of the greenhouse gases, which is a major reason for global warming. Global warming has a negative impact on the environment as it raises the sea level which as a result increases the pollution of the cost in many countries.[7] The global temperature of the earth's surface has been increased because of many industrial acts. Increasing carbon dioxide concentration to the double is expected to increase the surface's mean temperature by 1.5°C–4.5°C.[8] Ice caps' and glaciers' melting is a result of this temperature increase. 0.19 m is the increase in the sea level that has been indicated in the period of 1900–2010[9] and it is expected to rise to 0.52–0.98 by the end of this century.[10]

The literature review indicates the possibility of optimizing this process from different aspects like total production cost (TPC), AP, GWP, and product sales (PRS). The aim of this study is multiobjective optimization (MOO) of a selected production process. MOO is done for the minimization of AP along with TPC and minimization of GWP and maximization of PRS. The process is modeled using Aspen plus. Excel-based MOO (EMOO) with nondominated sorting genetic algorithm (NSGA-II) utilized to perform the optimization. Decision variables, such as flow rates of feed streams, operating pressure of some units, and operating temperature, were taken into consideration.

10.2   METHODOLOGY

10.2.1   PROCESS DESCRIPTION

In H_2SO_4 manufacturing process (Fig. 10.1), sulfur is firstly mixed with dry air in a burner to produce sulfur dioxide gas (SO_2). This gas is then sent to a reactor that consists of four passes for further oxidation to produce sulfur trioxide gas (SO_3). Then, SO_3 enters the absorption IPAT with water to produce the final product (H_2SO_4). Some of the product is used to remove water in a drying column and some of it is sent to absorption FAT to reduce SO_2 emissions.

10.2.2   PROBLEM FORMULATION

10.2.2.1   DECISION VARIABLES AND CONSTRAINTS

MOO of sulfuric acid process for the considered case is the simultaneous minimization of AP and TPC. Constraints for acid purity, SO_2 allowed emissions and air–water content were considered. The decision variables are reactor operating pressure (1.75–1.24 kPa vacuum), airflow rate (3396–3646 kmol/h), and water flow rate (4989–5898 kg/h). Additional decision variables are burner and FAT operating pressure, steam flow rate, and temperature.

10.2.2.2   OBJECTIVE FUNCTIONS

Case 1: Minimize AP and Minimize TPC
Case 2: Minimize GWP and Maximize PRS

Considering the acute impact of global warming and acidity, mini-mizing them is essential. To calculate both acidity, global warming, and acidification potential (GWP and AP respectively), the following equations are used[11]:

$$Normalized\ GWP = \frac{GWP}{Norm.ref._{GWP}} \qquad (10.1)$$

$$Weighted\ GWP = WF_{GWP} \times Normalized\ GWP \qquad (10.2)$$

where Normalized GWP is the normalized global warming potential for the considered substance, which is carbon dioxide in this case and $Norm.ref._{GWP}$ is the normalization reference for global warming. WF_{GWP} is the weighing factor for global warming.[11]

While for AP calculations, equations are[11]:

$$EF_i = \frac{n}{MW_i} \times 32.03 \qquad (10.3)$$

$$AP = \sum_i EF_i \times m_i \qquad (10.4)$$

MW_i is the molecular weight of the substance emitted, n is the number of hydrogen ions released, EF_i is the substance's equivalence factor.[11] In this sulfuric acid production plant, carbon dioxide is the only component contributing to the GWP, while sulfuric acid, sulfur trioxide, and sulfur dioxide are simultaneously participating in increasing AP.

10.3   RESULTS AND DISCUSSION

Case 1: As Figure 10.2a displays, the Pareto front for AP versus TPC has converged and there exists a conflict between these two objectives. Figures 10.2b, 10.2c, and 10.2d demonstrate the impact of the three passes operating pressure on TPC. It can be noticed that most of the results have converged close to the upper range limit (−1.24). Since the three passes are operating under vacuum, this trend is expected since the lower is the vacuum pressure, the higher is the operating cost, and hence TPC will also increase. The same thing is observed in their relationship with AP, Figures 10.2l, 10.2m, and 10.2n.

Similarly, the allocation of solutions for FAT pressure shows a slight preference for higher operating pressure with both objectives as Figures 10.2f and 10.2p show. The higher pressure will help in maximizing the absorption of acidic gases and, therefore, AP will decrease. For TPC,

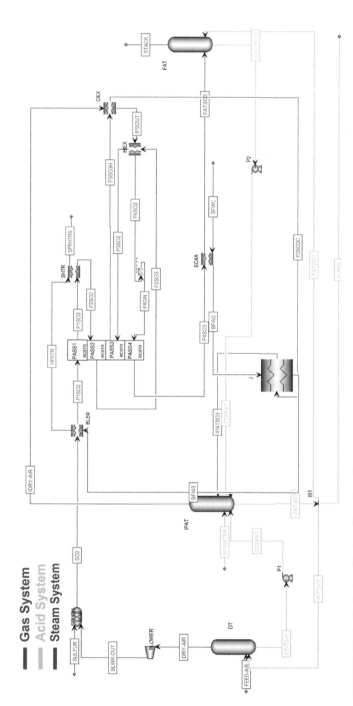

FIGURE 10.1 Sulfuric acid plant flow sheet.[12]

higher pressure works on decreasing the power needed in pump 2 (P2). By analyzing Figures 10.2e and 10.2o, it is seen that solutions are oriented in the area of the upper operating pressure range of IPAT, which can be elucidated by coupling this observation with the fact that the suitability of gases increases with increasing the pressure, which results in a less SO_x emissions and hence lower values for AP. The operating pressure of the burner converged in the lower range of the decision variable (Figs. 10.2g and 10.2q).

Figure 10.2h and r show the impact of changing the steam flow rate along with other variables in the specified range on AP and TPC. It is noticed that results have converged in the lower range of the decision variable. Lower steam flow rates will minimize TPC and will keep the desired temperature in the reactors for higher conversion of SO_2, which means that less of this gas is emitted to the atmosphere. On the other hand, the water flow rate results are in the middle of the decision variable range (Figs. 10.2i and 10.2s). Increasing this variable has the power of reducing AP as it will react with SO_3 to produce more acid, but at the same time, it will increase the production cost. To balance this relationship, the convergence trend is reasonable.

Airflow rate is affecting the considered objectives as seen in Figures 10.2j and 10.2t. The increase in air flow rate will directly raise the power needed in the compressor which will maximize TPC. On the other hand, higher airflow rates are needed for higher conversion in the burner and the reactor, which will minimize AP. In contrast to what is observed for airflow rate and most of the other variables, varying the temperature of steam in the specified range had no distinguishable impact on AP and TPC (Figs. 10.2k and 10.2u).

Case 2: GWP and PRS case is optimized and the Pareto front along with all decision variable relations are shown in Figure 10.3. Figures 10.3b, 10.5c, and 10.3d elucidate the suzerainty of passes operating pressure on PRS. As appears, the results are concentrated in the upper pressure bound for pass1 (more than 83%) and pass2, as seen in Figures 10.3b and 10.3c, respectively. Since the target is to maximize PRS, which is highly dependent on the amount of SO_3 gas, and since this gas is formed in theses passes, then higher pressure is favored to give the desired gas which will increase acid production. On the other hand, according to the increase of SO_3 in concentration in gas streams, the CO_2 concentration will decrease, so GWP is minimized (Figs. 10.3m and 10.3n).

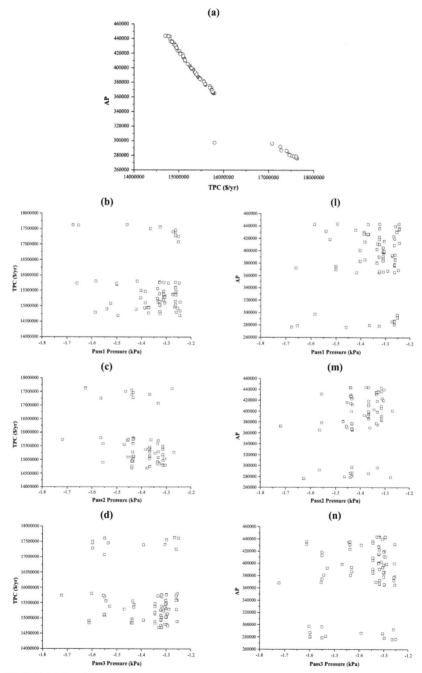

FIGURE 10.2 *(Continued)*

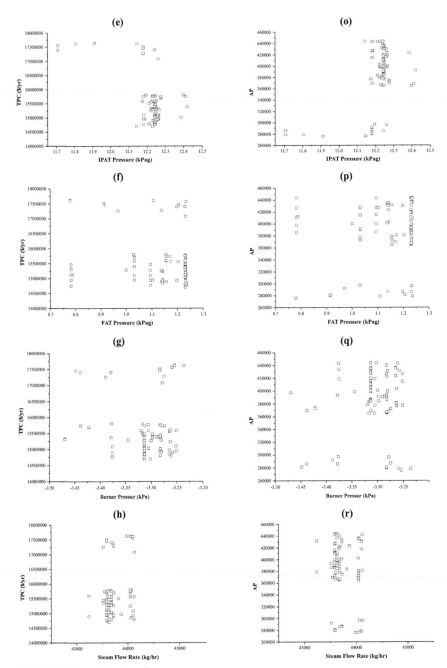

FIGURE 10.2   *(Continued)*

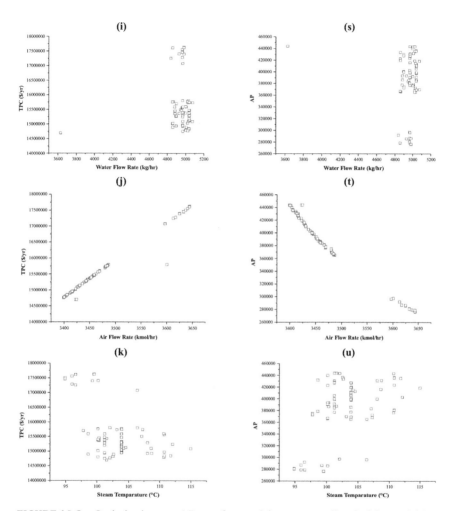

FIGURE 10.2 Optimization case 1 Pareto front and the corresponding decision variables.

For pass3, as shown in Figures 10.3d and 10.3o for PRS and GWP, respectively, the results are mostly converged to the lower bound of the decision variable (−1.692 to −1.493 kPa) on the contrary to former two passes. IPAT pressure tended to converge in the upper range as seen in Figures 10.3e and 10.3p. As justified before, the gas is more soluble in water at a higher pressure so the conversion of the reaction responsible for acid formation will increase to give more products to increase sales (Fig. 10.3e). Also, the absorption of CO_2 rise with higher pressure values, which

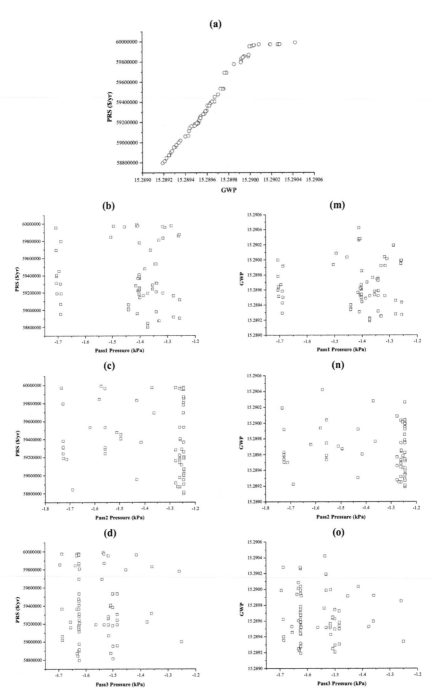

FIGURE 10.3   *(Continued)*

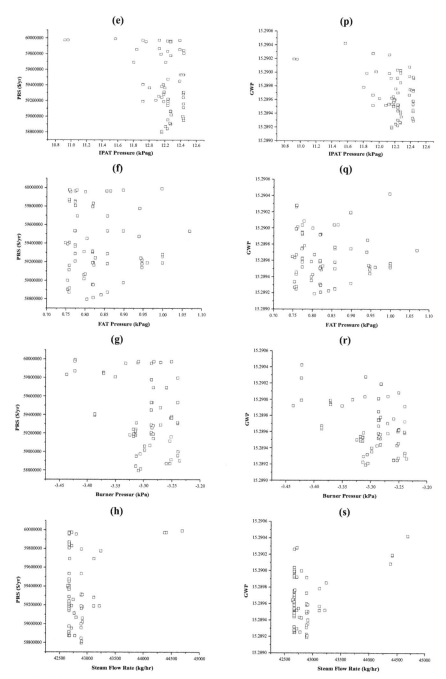

FIGURE 10.3 *(Continued)*

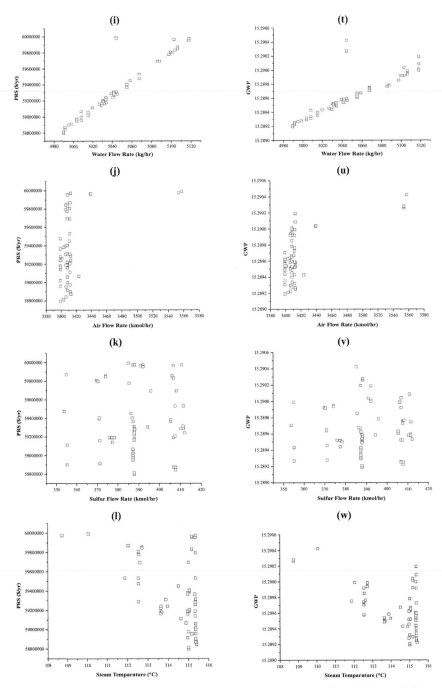

FIGURE 10.3   Optimization case 2 Pareto front and the corresponding decision variables.

justify the trend in Figure 10.3p. Likewise, FAT operating pressure follows the same trend as shown in Figures 10.3f and 10.3q for the same reasons.

Figures 10.3g and 10.3r display the impact of burner pressure on objectives. It is noticed that most solutions converged to the upper bound of the decision variable (85% of solutions are above −3.334 kPa), which could be needed to achieve higher oxidation rate for sulfur, which consequently gives higher product rate to increase PRS (Fig. 10.3g) and minimize GWP as well as CO_2 formation rate decreases (Fig. 10.3r).

The low flow rate of steam is essential to achieve the desired conversion in the multipass reactor to maximize PRS and minimize GWP as described earlier (Figs. 10.3h and 10.3s). Airflow rate can also justify the less amount needed of steam for cooling as more than 94% of solutions tend to be in its lower range area, >3442.5 kmol/h, so less amount of gas needs to be cooled (Figs. 10.3j and 10.3u). In the same regard, less air means that less CO_2 is produced.

Figures 10.3i and 10.3t show the effect of the water flow rate. Mounting the rate of SO_3 production means that more water is required to convert it to H_2SO_4 and hence increase PRS (Fig. 3i). On the other hand, converting more SO_3 to H_2SO_4 leads to rise in the concentration of CO_2 leaving the IPAT column to the FAT column, and this GWP will increase (Fig. 10.3t).

The flow rate of sulfur is not of significant impact on PRS and GWP as Figures 10.3k and 10.3v exhibit. Solutions are scattered on the entire range of sulfur flow rate. High values of PRS were achieved at the high and low temperatures of steam (Fig. 10.3l), but GWP tends to become lesser at a higher steam temperature as shown in Figure 10.3w. This tendency for higher steam temperature is needed for the gas to achieve the needed reaction temperature for the best conversion in the three passes to achieve more product and increase sales.

10.4 CONCLUSION

MOO was done using EMOO with NSGA-II for the considered industrial cases. Various decision variables and constraints were taken into consideration. It was concluded that, for case 1, the flow rates of water, steam, and air highly affected the obtained results. As expected, TPC increased with air flow rate while the opposite trend was seen with AP. Steam flow rate converged to the lower decision variable bound, whereas water flow rate converged to values that will satisfy both objectives (approximately the middle of the rang). In general, the operating pressure of all units converged to the upper limit of their decision variables range to meet the desired selected objectives. For case 2, the water flow rate was the decision variable with the highest influence on the considered case because it caused the conflict between both objectives (minimization of GWP and maximization of PRS), as both objectives increase with increasing the water flow rate. In addition, most of the solutions converged at the lower bound of airflow rate. This work presents a deeper understanding of the process that will help related process engineers to develop their plants.

KEYWORDS

- multiobjective optimization
- evolutionary algorithms
- H_2SO_4
- global warming potential
- economics

REFERENCES

1. Ciobanu, T. Sulfuric Acid Production. In *Pollution Control in Fertilizer Production.* Marcel Dekker: NY; 1994; p. 161–186.
2. Ashar, N. G.; Golwalkar, K. R. Processes of Manufacture of Sulfuric Acid. In *A Practical Guide to the Manufacture of Sulfuric Acid, Oleums, and Sulfonating Agents.* Springer; 2013; p. 9–30.
3. Kiss, A. A.; Bildea, C. S.; Verheijen, P. J. T. Optimization Studies in Sulfuric Acid Production. In *16th European Symposium on Computer Aided Process Engineering and 9th International Symposium on Process Systems Engineering.* Elsevier; 2006; p. 737–742.
4. Telang, K. S. *Advanced Process Analysis System.* Louisiana State University: Baton Rouge; 1998.
5. Bakhshipour, Z., et al. Effect of Acid Rain on Geotechnical Properties of Residual Soils. *Soils Found.* **2016,** *56* (6), 1008–1020.
6. Santamarina, J. C., et al. Specific Surface: Determination and Relevance. *Canadian Geotec. J.* **2002,** *39* (1), 233–241.

7. Stocker, T. F., et al. Climate Change 2013: The Physical Science Basis. Contribution of Working Group I to the Fifth Assessment Report of the Intergovernmental Panel on Climate Change. Cambridge Univ. Press: Cambridge, UK, and New York; 2013; pp. 1535.

8. Lindsey, R. *How Much will Earth Warm if Carbon Dioxide Doubles Pre-industrial Levels.* National Oceanic and Atmospheric Administration, USA, 2014.

9. Church, J. A., et al. Sea-Level Rise by 2100. *Science* **2013,** *342* (6165), 1445–1445.

10. DeConto, R. M.; Pollard, D. Contribution of Antarctica to Past and Future Sea-level Rise. *Nature* **2016,** *531* (7596), 591–597.

11. Stranddorf, H.; Hoffmann, L.; Schmidt, A. *Impact categories, normalisation and weighting in LCA (Påvirkningskategorier, normalisering og vægtning i LCA–in Danish), Environmental News No. 77, The Danish Ministry of the Environment.* Environmental Protection Agency, Copenhagen, 2004.

12. Aspen Technology, I., *Aspen Plus Sulfuric Acid Model.* In *Technical Report,* 2008.

CHAPTER 11

Simultaneous Optimization Aspects in Industrial Gas Sweetening Process for Sustainable Development

DEBASISH TIKADAR[1, 2, 3,] CHANDAN GURIA[2], and ASHISH M. GUJARATHI[1*]

[1]Department of Petroleum and Chemical Engineering, Sultan Qaboos University, Muscat 123, Oman

[2]Department of Petroleum Engineering, Indian Institute of Technology (Indian School of Mines), Dhanbad 826004, India

[3]Worley Oman Engineering LLC, Muscat 133, Oman

*Corresponding author. E-mail: ashishg@squ.edu.om

ABSTRACT

Natural gas is one of the leading source of energy. Raw natural gas mainly contains methane along with trace amount of other hydrocarbons and some impurities like H_2S and CO_2, which are very toxic and hazardous gas for leaving being. This gas is being treated in industry to remove the impurities before it is sent to the various industrial and domestic users. Due to instability of crude oil price in the global market and for the sustainable development, it is now important to optimize the existing gas sweetening plant with respect to economic and environmental criteria. A model of gas sweetening process is developed in Aspen Hysis V10 process simulation software and validated by plant data. Multi-objective optimization study is carried out by using I-MODE algorithm. Excel-based VBA codes are used to interlink Hysys model and I-MODE code. In this paper the effects of different independent variables such as lean amine temperature, feed gas temperature, regenerator feed temperature, and pressure are considered with their respective bounds and optimization study is carried out for two different conflicting objectives like damage index and CO_2 removal, and so forth. The results are discussed and analyzed with their Pareto optimal set of solutions.

11.1 INTRODUCTION

Due to the utilization of huge amount of energy in the development of technology, potential environmental threats such as global warming, depletion of ozone layer, acidification, inherent safety, and so forth, are increasing day-by-day. The chemical industries are one of the areas, which are responsible for this issue. Clean technology and industrial ecology are the major consideration for process selection, basic engineering, and detailed design of a chemical plant. Most of the companies are always compromising environmental and safety-oriented objectives to get more profitable business. It is now essential to resolve this global issue with the involvement of chemical engineering with other disciplines including science, economics, environmental, and toxicology. Hence, the leading chemical companies and industry associations are taking initiatives for sustainable development of existing chemical plant. Scientists and engineers are also working for a potential development of chemical industries by modifying the existing technology of replacing with new energy-efficient process technology. The sustainable and innovative process/product development can be accelerated through the collaborative work together with scientists, engineers, and public/private companies.

Optimization is a mathematical technique to solve the quantitative problems in the area of science, engineering, and economics. It deals with maximization or minimization of real objective function by changing the decision variables. Optimization of process plant is one of the major area in the chemical engineering field. Usually, process engineers need to optimize the process plant in terms two different aspects, that is, operation optimization and design optimization to improve the costs and profitability of the project, operation performance of the plant, technical safety and reliability of the plant, and environmental stewardship. Multi-objective optimization of many process engineering problems can be formulated and solved by using the numerical methods under the field of optimization. The optimization of a chemical process plant is more challenging as the process engineering application. These problems are often complicated as several objective functions are involved. Existence of trade-off between the two-objective functions makes the optimization problems more complex. It involves constrain handling and involves both continuous and discrete decision variables.

A multi-objective optimization problem can be stated as follows[1]:

$$\text{Maximize/Minimize } f_m(x), \text{ where, } m = 1, 2, \ldots, M \qquad (11.1)$$
subject to

$$g_j(x) \geq 0, \qquad\qquad j = 1, 2, \ldots, J \qquad (11.2)$$
$$h_k(x) = 0, \qquad\qquad k = 1, 2, \ldots, K \qquad (11.3)$$
$$x_i^{(L)} \leq x_i \leq x_i^{(u)}, \qquad\qquad i = 1, 2, \ldots, n \qquad (11.4)$$

where, m is the number of objective functions to be simultaneously optimized, x is the vector of n decision variables (continuous and/or discontinuous) with lower $x_i^{(L)}$ and upper $x_i^{(u)}$ bounds, J and K are the number of inequality (g) and equality (h) constraints, respectively.

Recent past many researchers have solved process engineering problems by using multi-objective optimization technique. Multi-objective optimization of an industrial hydrocracking unit is studied by Bhutani et al.[2] to identify the optimal solutions for reactor kinetic model. They solve two-objective optimization by using real-coded elitist NSGA and analyzed the results along with Pareto optimum set of solutions. Gujarathi and Babu[3] developed H-MODE algorithm and optimized adiabatic styrene reactor. Four study cases were formulated under this study and discussed the optimization results. Multi-objective optimization study was conducted by Gujarathi and Babu[4] for industrial styrene reactor. They considered productivity, selectivity, and yield of styrene reactor as an objective function. Gujarathi and Babu[5] also carried out comparative study of Elitist-MODE and MODE-III for different industrial processes. Improved multi-objective differential evolution algorithm is developed by Sharma and Rangaiah and they set a termination criterion for optimizing a processes plant.[6] They also carried out multi-objective optimization of two different engineering problems for alkylation process and fermentation process by using I-MODE algorithm to obtain global optimal solutions and its application to process engineering problems.[7] Guria et al.[8] carried out optimization of reverse osmosis desalination process unit. They used binary coded elitist non-dominated sorting genetic algorithm (NSGA-II) to obtain the solutions. Guria et al.[9] also studied froth flotation circuits for mineral processing by using the Jumping gene adaptation of genetic algorithm. They obtained Pareto optimal non-dominated set of solutions for maximization of the throughput and minimization of cost. Industrial naphtha catalytic reforming process is optimized by Hou et al.[10] They used NSGA algorithm to solve the problem and obtained Pareto front and different set of decision variables.

Natural gas is one of the major sources of energy that is available from oil wells and gas wells. Raw crude is obtained from wells to inlet separator via flow line and pipeline where gas is separated from oil and water. The raw gas contain some impurities like water, hydrogen sulfide (H_2S), carbon di-oxide (CO_2), inert gases like nitrogen, and so forth. These impurities may be a cause of corrosion, erosion to the transport pipeline and other equipment, lowering the heating value, and H_2S is also harmful and forms a part of toxic impurities.[11] The sales price of natural gas mostly depends on its heating value. Hence, it is essential to separate the impurities from these gases before further processing. The different gas treatment and processing units such as acid gas removal unit, dehydration unit, and finally separation/fractionation unit for different products are shown in Figure 11.1. Gross crude with associated gas from different wells are collected through flow line and sent to inlet separator to separate oil, water, and gas. This gas is again sent to the gas sweetening unit to separate H_2S and CO_2. Acid gas removal process is chosen to conduct the optimization study.

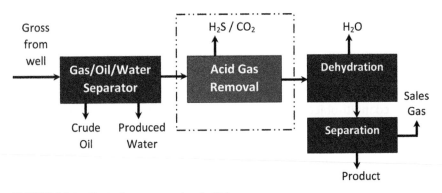

FIGURE 11.1   Typical gas processing facilities.

Different methods are established in the recent past for recovering sulfur compounds from natural gas. The most established process for onshore use is the amine processes. Alkanolamine originally applied by Bottoms since 1930.[12] Thereafter, amine processes are developed and have become the most widely used solvents for the acid gas removal from natural gas. Al-Lagtah et al.[13] carried out optimization and performance improvement of gas sweetening for Lekhwair plant using Aspen Hysys. They reduced 50% of operating cost by modifying the split loop configuration of Lekhwair plant configuration. Borhani et al.[14] studied modeling and simulation of simultaneous removal of CO_2 and H_2S from natural gas by using MDEA solution. Aqueous amine solution and liquid propane mixture is used by Carroll et al.[15] to study the H_2S distribution in this solvent solution. Ghanbarabadi and Gohari[16] studied the optimization of MDEA-based absorption tower and developed the effect of amine concentration on the process plant performance. Behroozsarand and Zamaniyan[17] optimized industrial synthesis gas sweetening plant in GTL process by using NSGA-II. They considered net energy consumption, amine circulation, and CO_2 recovery as the problem object function.

This study shows that, most of the work is done for acid gas treatment unit based on modeling, simulation, and sensitivity analysis. Thorough multi-objective optimization study for industrial gas sweetening process is not yet performed for the reasons of safety and process-related objectives. Hence, optimization of natural gas sweetening process is considered for this study.

11.2 PROBLEM FORMULATION

Natural gas sweetening process unit is selected to study the multi-objective function optimization by using I-MODE algorithm. A traditional method of natural gas treatment for sweetening unit is alkanol amine process of natural gas.[12] Three different amines are used in gas sweetening; namely, monoethanolamine (MEA), diethanolamine (DEA), and methyldiethanolamine (MDEA). Among them, MDEA-based gas sweetening process is considered under this optimization study. Absorption and regeneration processes are involved and the chemical reactions are reversible in nature. Exothermic reactions take place in absorption column and endothermic reactions take place in regeneration column. The following physical and chemical reactions are involved in the gas sweetening process.[18,19]

Physical equilibrium:

$$H_2S \text{ (g)} \rightleftharpoons H_2S \text{ (aq)} \tag{11.5}$$

$$CO_2 \text{ (aq)} \rightleftharpoons CO_2 \text{ (g)} \tag{11.6}$$

$$CH_4 \text{ (aq)} \rightleftharpoons CH_4 \text{ (g)} \tag{11.7}$$

$$H_2O \text{ (1)} \rightleftharpoons H_2O \text{ (g)} \tag{11.8}$$

$$MDEA \text{ (aq)} \rightleftharpoons MDEA \text{ (g)} \tag{11.9}$$

Chemical reactions:

Ionization of water:

$$H_2O \text{ (1)} \rightleftharpoons OH^- \text{ (aq)} + H^+ \text{ (aq)} \tag{11.10}$$

Protonation of MDEA:

$$MDEA \text{ (aq)} + H_2O \text{ (1)} \rightleftharpoons MDEAH^+ \text{ (aq)} + OH^- \text{ (aq)} \tag{11.11}$$

Dissociation of carbon dioxide:

$$CO_2 \text{(aq)} + OH^- \text{ (aq)} \rightleftharpoons HCO_3^- \text{ (aq)} \tag{11.12}$$

Dissociation of bicarbonate ion:

$$HCO_3^- \text{ (aq)} + OH^- \text{ (aq)} \rightleftharpoons CO_3^= \text{ (aq)} + H_2O(1) \tag{11.13}$$

Hydrogen sulfide dissociation:

$$H_2S \text{ (aq)} \rightleftharpoons H^+ \text{ (aq)} + HS^- \text{ (aq)} \tag{11.14}$$

Bisulfide ion dissociation:

$$HS^- \text{ (aq)} \rightleftharpoons H^+ \text{ (aq)} + S^= \text{ (aq)} \tag{11.15}$$

In general, natural gas obtained from gas/oil wells contains H_2S and CO_2, which is above the gas pipeline specification. The maximum allowable limit of H_2S and CO_2 content is 4 ppm and 2% mol.[12] The feed gas that is used in this study contains 200 ppm H_2S and 0.26% mol of CO_2.[13] The typical gas sweetening process schematic diagram is shown in Figure 11.2. It includes absorption column, regeneration column, heat exchanger, cooler, re-boiler, pump, surge tank, flash tank, reflux drum, and so forth, like major equip*ment.*

Three-objective functions; namely, carbon di-oxide removal, sulfur di-oxide removal, and damage index (DI) are considered for natural gas sweetening process optimization study. Seven decision variables such as feed gas pressure, feed gas temperature, lean amine pressure, lean amine temperature, regenerator feed pressure, regenerator feed temperature, and

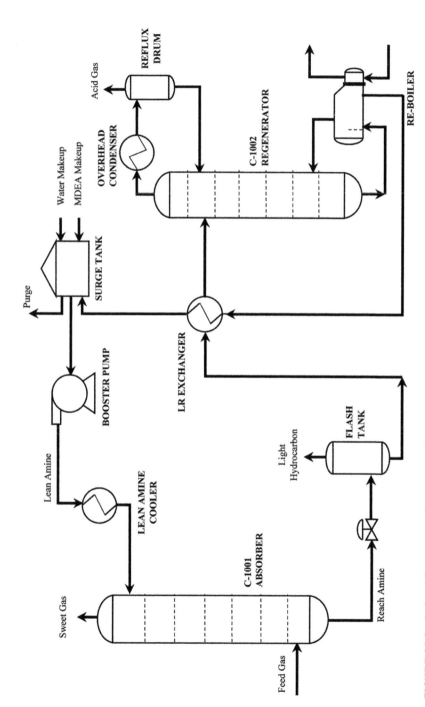

FIGURE 11.2 Industrial gas sweetening process scheme.

feed gas molar flow are used for these cases. First two-objective functions are related to process operation oriented whereas third or last one is inherent safety-oriented objective.

Carbon di-oxide (CO_2) removal is defined as the percentage of CO_2 removed from feed gas

$$CO_2 \text{ Removal } (\%), J1 = \frac{m_{1,co_2} - m_{2,co_2}}{m_{1,co_2}} \times 100\% \qquad (11.16)$$

Where, m_{1,co_2} is the feed gas CO_2 mass flow rate, kg/h, and m_{2,co_2} is the sweet gas CO_2 mass flow rate, Kg/h. The objective is to minimize export gas CO_2 content, which is less than 2% mole, that is, maximize the CO_2 removal.

Hydrogen sulfide (H_2S) removal is defined as the percentage of H_2S removed from feed gas

$$H_2S \text{ Removal } (\%), J2 \frac{m_{1,h_2s} - m_{2,h_2s}}{m_{1,h_2s}} \times 100\% \qquad (11.17)$$

Where, m_{1,h_2s} is the feed gas H_2S mass flow rate, kg/h, and m_{2,h_2s} is the sweet gas H_2S mass flow rate, kg/h. The objective is to minimize export gas H_2S content, which is less than 4 ppm, that is, maximize the H_2S removal.

Khan and Amyotte[20–21] studied the inherent safety-related objectives and developed the integrated inherent safety index (I2SI) to quantify the inherent safety indices. I2SI is inversely proportional to damage index (DI) and hence, to design inherently safer process, it is required to minimize DI, which is equivalent to maximization of I2SI. In this study, DI is taken as an inherent safety objective for gas sweetening process optimization, which can be calculated as:

$$DI, J3 = \sqrt{(total\ FEDI)^2 + (total\ TDI)^2} \qquad (11.18)$$

Where, FEDI and TDI are estimated based on FEDR and TDR, which are calculated using the SWeHI approach.[22] The main objective of this study is to increase I2SI by minimizing DI.

Two multi-objective optimization problems for natural gas sweetening plant are formulated under this study. One bi-objective optimization and other is three-objective optimization problem. It is observed that a trade-off exists between the percentage of CO_2 removal and DI. If we increase CO_2 removal DI will increase. However, it is desired to reduce DI. This is obvious that the CO_2 removal from raw natural gas is a costly process, and it involves high pressure in absorption column as well as high pressure in regeneration column. Hence, with increase in CO_2 removal DI increases. Similarly, three-objective optimization problem is maximization of percentage of CO_2 removal, maximization of percentage of H_2S removal, and minimization of DI. The objective functions, decision variables with their bounds, and constraints for this optimization study are given as:

Objective functions:

Problem 1: Max. $J1 = (\dfrac{m_{1,co_2} - m_{2,co_2}}{m_{1,co_2}} \times 100\%)$,

Min. $J3 = \sqrt{(total\ FEDI)^2 + (total\ TDI)^2}$,

Problem 2: Max. $J1 = (\dfrac{m_{1,co_2} - m_{2,co_2}}{m_{1,co_2}} \times 100\%)$,

Max. $J2 = (\dfrac{m_{1,h_2s} - m_{2,h_2s}}{m_{1,h_2s}} \times 100\%)$, Min. $J3 =$

Decision variables:

Feed gas pressure, kPa	$5700 \leq Pf \leq 9000$
Feed gas temperature, °C	$40 \leq Tf \leq 60$
Lean amine pressure, kPa	$5000 \leq Pla \leq 9000$
Lean amine temperature, °C	$45 \leq Tla \leq 65$
Regenerator feed pressure, kPa	$200 \leq Prf \leq 280$
Regenerator feed temperature, °C	$90 \leq Trf \leq 125$
Feed gas molar flow, kmol/h	$4000 \leq Ff \leq 5000$

Constraints:

Sweet gas H_2S content, % mol	≤ 0.001
Sweet gas CO_2 content, % mol	≤ 2

Improved multi-objective differential evaluation algorithm (I-MODE) is employed to optimize natural gas sweetening process. Sharma and Rangaiah (2013) developed I-MODE by including taboo list with MODE algorithm and introduced termination criterion for three chemical engineering problems; namely, fermentation, alkylation, and Williams-Otto process. The parameters used for optimization study by I-MODE algorithm are Cross Over Probability: 0.9, Mutation Probability: 0.02, Tabu List: 60, Tabu Radius: 0.01, VAR-limit (GD): 0.0003, VAR-limit (SP): 0.1, SSTC:

λ1: 0.1, SSTC: λ2: 0.1, SSTC: λ3: 0.1, SSTC: Rcrit: 0.9, Population Size: 120, Max. No. of Generations for two-objective optimization case: 120 and Max. No. of Generations for three-objective optimization case: 110.

Aspen Hysys is a well-known process simulator software, which is used by process design engineers worldwide. A Hysys model is developed and validated this model with natural gas sweetening plant data[23] to carry out the gas sweetening process optimization. It is seen that the results in close agreement to the best published results available in the literature.[13] Excel-based EMOO program is used to carry out the optimization study, which is basically interlinking software among Excel, Visual Basic for Applications (VBA), and Hysys. The I-MODE algorithm code is implemented by using VBA in Microsoft Excel 2016. The I-MODE binary code generates different sets of decision variables and these data are transferred to Hysys. In Excel spreadsheet all the objective functions and constraints are calculated using the data generated from Hysys once the simulation is converged for one set of random decision variables. These calculated objective functions and constraints are used to optimization analysis by I-MODE. Then I-MODE generates another set of solutions for similar objective functions and constraints, and these procedures and the iterative calculations are repeated until the specified criterion is satisfied.[24]

11.3   RESULTS AND DISCUSSION

In this study, the results of two-objective optimization case-1 and three-objective optimization case-2 are discussed. Both problems are analyzed with their Pareto optimal set of solutions and compared. Effects of seven decision variables on objective functions are clearly investigated. The trade-off between two-objective functions are analyzed and discussed in the later section.

11.3.1   TWO-OBJECTIVE OPTIMIZATION

Problem-1: Maximization of CO_2 removal and minimization of DI

Simultaneously optimization of CO_2 removal (eq 11.16) and Damage Index (eq 11.18) are considered. Figure 11.3a shows the Pareto optimal set of solutions for the objective functions of CO_2 removal and DI for 120 numbers of generation. Here, with the increase in CO_2 removal the DI increases, which is against the requirement of this objective function DI. Any of the one objective, either CO_2 removal or DI, needs to be sacrificed to get better result of another one. Feed gas pressure is converged at its lower bound of 5700 kPa (Fig. 11.3b). It is seen that, in Figure 11.3c feed gas temperature is converged at its upper bounds of 59°C. Preliminary analysis shows that lean amine pressure is scattered between 5200 kPa to 8000 kPa. However, it is seen that, most of the solution points remain in the range of 5200 kPa to 6300 kPa, which falls within its lower bounds (Fig. 11.3d). Figures 11.3e and 11.3f show variation of lean amine temperature and regeneration feed pressure with CO_2 removal. Here, lean amine temperature is nearly constant (at 62°C), while regeneration feed pressure is scattered between 202°C and 272°C. But 82% data point of regeneration feed pressure fall under the range of 202°C to 210°C. Probably this may be the cause of the other objective function (DI), which always demands low pressure and temperature for this optimization case. The regenerator feed temperature converged between 112°C and 117°C (Fig. 11.3g). It is seen that the feed gas molar flow is converged at its lower bound (4000 kmol/h) (Fig. 11.3h). This is due to the effect of DI, that is, due to lowering of flow rate to maintain less DI. DI and CO_2 removal is conflicting in nature and hence, to optimize these two objectives the decision maker can study the Pareto front and select the decision variable to get the optimize results.

11.3.2   THREE-OBJECTIVE OPTIMIZATION

Problem-2: Maximization of CO_2 removal, maximization of H_2S removal, and minimization of DI.

The trade-offs between two objectives, that is, CO_2 removal and DI are analyzed in Problem-1. Similarly, three-objective optimization study is carried out for the objective functions of percentage of CO_2 removal, percentage of H_2S removal, and DI. For the industrial application three-objective optimization study is more relevant than two-objective cases. In this case, percentage of CO_2 removal and percentage of H_2S removal are to be maximized while minimizing DI. It is observed that three-objective optimization becomes more complicated and it takes more time to converge the simulation. Three-objective optimization results are plotted in 3D as shown in Figure 11.4, and it is seen that among these

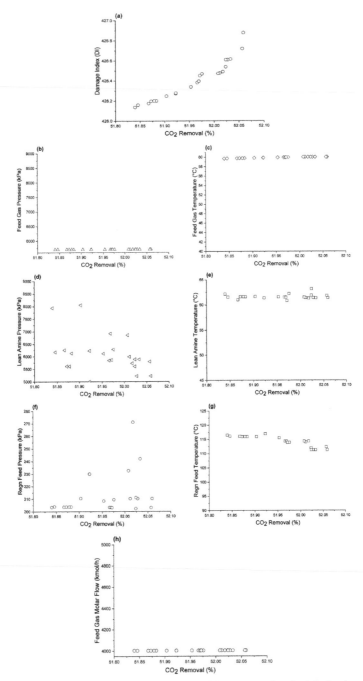

FIGURE 11.3   MOO results for maximization of CO$_2$ removal and minimization of DI.

three-objective function, any of the two-objective functions are conflicting in nature. The comparative study between two-objective optimization and three-objective optimization results are shown in Figure 11.5. It is noticed that the plot between CO_2 removal and DI for two- and three-objective optimization cases in Figures 11.5a and 11.5b show the solutions of the two-objective case lie on the perimeter of the plot formed by the best solutions of the three-objective study. Figure 11.5b is the extended version of the Figure 11.5a, that is, the common Pareto front area of two- and three-objective cases. It is also observed that, the three-objective optimal points go away from the two-objective solutions. This is because of the third objective interferences, which also carries equal weightage to the other two objectives.

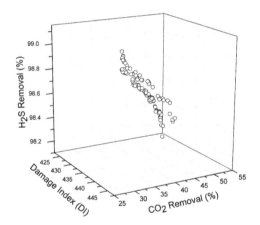

FIGURE 11.4 MOO results for maximization of CO_2 removal maximization of H_2S removal and minimization of DI.

11.4 CONCLUSION

Operation optimization study is carried out to optimize process and safety objective for gas sweetening process. In this study, DI, H_2S removal, and CO_2 removal are considered as objective function, which affect the performance of the plant. Seven decision variables with their respective upper and lower bounds are used to solve the two-objective and three-objective optimization problem by using I-MODE. The results are analyzed and explained in the results and discussion section and the overall findings are mentioned as follows:

- With increase in CO_2 removal DI increases.
- CO_2 removal improved by lowering the regenerator feed temperature.
- CO_2 removal better for the lean amine pressure of 6000 kPa.
- Best operating condition can be selected for the maximum CO_2 removal of 51.7% to 52.1%.
- The two-objective optimization study provides better solutions than that of three-objective optimization problem. However, three objective optimization results are more rational.

Both the cases of two and three objective problems are optimized with the specified ranges of seven decision variables and the optimal solutions are explained qualitatively. Plant operator and process engineer can select the preferred solution from the Pareto front based on the economic and environmental consideration.

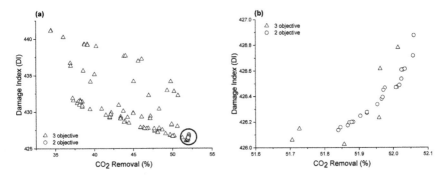

FIGURE 11.5   Comparison of Pareto-optimal solutions for the two-objective and three-objective optimization study at generation 210 for the gas sweetening process.

KEYWORDS

- **optimization**
- **I-MODE**
- **industrial gas sweetening**
- **MDEA Hysys**

REFERENCES

1. Deb, K. Multi-Objective Optimization Using Evolutionary Algorithms, 1st ed. John Wiley & Sons Ltd.: Chichester, 2001.
2. Bhutani, N.; Ray, A. K.; Rangaiah, G. P. Modeling, Simulation, and Multi-Objective Optimization of an Industrial Hydrocracking Unit. *Ind. Eng. Chem. Res.* **2006,** *45* (4), 1354–1372.
3. Gujarathi, A. M.; Babu, B. V. Optimization of Adiabatic Styrene Reactor: A Hybrid Multi-Objective Differential Evolution (H-MODE) Approach. *Ind. Eng. Chem. Res.* **2009,** *48* (24), 11115–11132.
4. Gujarathi, A. M.; Babu, B. V. Multi-objective Optimization of Industrial Styrene Reactor: Adiabatic and Pseudo-isothermal Operation. *Chem. Eng. Sci.* **2010,** *65* (6), 2009–2026.
5. Gujarathi, A. M.; Babu, B. V. Multi-Objective Optimization of Industrial Processes using Elitist Multi-Objective Differential Evolution (Elitist-MODE). *Mater. Manuf. Processes.* **2011,** *26* (3), 455–463.
6. Sharma, S.; Rangaiah, G. P. An Improved Multi-Objective Differential Evolution with a Termination Criterion for Optimizing Chemical Processes. *Comput. Chem. Eng.* **2013,** *56*, 142–154.
7. Sharma, S.; Rangaiah, G. P. Hybrid Approach for Multiobjective Optimization and Its Application to Process Engineering Problems. In *Applications of Metaheuristics in Process Engineering*; Valadi, J. and Siarry, P. Eds.; Springer International: Switzerland, 2014; pp. 123–444.
8. Guria, C.; Bhattacharya, P. K.; Gupta, S. K. Multi-Objective Optimization of Reverse Osmosis Desalination Units Using Different Adaptations of the Non-dominated Sorting Genetic Algorithm (NSGA). *Comput. Chem. Eng.* **2005a,** *29* (9), 1977–1995.
9. Guria, C.; Varma, M.; Mehrotra, S. P.; Gupta, S. K. Multi-Objective Optimal Synthesis and Design of Froth Flotation Circuits for Mineral Processing, Using the Jumping Gene Adaptation of Genetic Algorithm. *Ind. Eng. Chem. Res.* **2005b,** *44* (8), 2621–2633.
10. Hou, W.; Su, H.; Mu, S.; Chu, J. Multiobjective optimization of the industrial naphtha catalytic reforming process. *Chin. J. Chem. Eng.* **2007,** *15* (1), 75–80.
11. Raymond, M. S.; Leffler, W. L. Oil & Gas Production in Nontechnical Language. Pennwell: Tulsa, Oklahoma, 2006.
12. GPSA, *GPSA Engineering Data Book*, 11th ed. Gas Processors and Suppliers Association: Tulsa (OK, USA), 1998.
13. Al-Lagtah N. M. A.; Al-Habsi S.; Onaizi S. A. Optimization and Performance Improvement of Lekhwair Natural Gas Sweetening Plant using Aspen Hysys. *J. Nat. Gas Sci. Eng.* **2015,** *26*, 367–381.
14. Borhani, T. N. G.; Afkhamipour, M.; Azarpour, A.; Akbari, V.; Emadi, S. H.; Manan, Z. A. Modeling Study on CO_2 and H_2S Simultaneous Removal using MDEA Solution. *J. Ind. Eng. Chem.* **2016,** *34*, 344–355.
15. Carroll, J. J.; Jou, F.; Mather, A. E.; Otto, F. D. The Distribution of Hydrogen Sulfide between an Aqueous Amine Solution and Liquid Propane. *Fluid Phase Equilibr.* **1993,** *82*, 183–190.

16. Ghanbarabadi, H.; Gohari, F. K. Z. Optimization of MDEA Concentration in Flow of Input Solvent to the Absorption Tower and Its Effect on the Performance of Other Processing Facilities of Gas Treatment Unit in Sarakhs Refinery. *J. Nat. Gas Sci. Eng.* **2014**, *20*, 208–213.

17. Behroozsarand, A.; Zamaniyan, A.; Multiobjective Optimization Scheme for Industrial Synthesis Gas Sweetening Plant in GTL Process. *J. Nat. gas Chem.* **2011**, *20* (1), 99–109.

18. Sadegh, N.; Stenby, E. H.; Thomsen, K. Thermodynamic Modelling of Acid Gas Removal from Natural Gas using the Extended UNIQUAC Model. *Fluid Phase Equilibr.* **2017**, *442*, 38–43.

19. Niu, M. W.; Rangaiah, G. P. Retrofitting Amine Absorption Process for Natural Gas Sweetening via Hybridization with Membrane Separation. *Int. J. Greenh. Gas Con.* **2014**, *29*, 221–230.

20. Khan, F. I.; Amyotte, P. R. Integrated Inherent Safety Index (I2SI): A Tool for Inherent Safety Evaluation. *Process Saf. Prog.* **2004**, *23* (2), 136–148.

21. Khan, F. I.; Amyotte, P. R. I2SI: A Comprehensive Quantitative Tool for Inherent Safety and Cost Evaluation. *J. Loss Prevent. Proc.* **2005**, *18*, 310–326.

22. Rangaiah, G.P.; A. Bonilla-Petriciolet, A. (Editors), *Multi-Objective Optimization in Chemical Engineering: Developments and Applications*, John Wiley: New Jersey, 2013.

23. Tikadar, D.; Guria, C.; Gujarathi, A. M. Multi-Objective Optimization of an Industrial Gas-Sweetening Unit using Process, Economic and Environmental Criteria. *J. Nat. Gas Sc. Eng.* **2019**, (Communicated on 30 March 2019).

24. Sharma, S.; Chua, Y. C.; Rangaiah, G. P.; Economic and Environmental Criteria and Trade-Offs for Recovery Processes. *Mater. Manuf. Processes.* **2011**, *26* (3), 431–445.

CHAPTER 12

A DFT Study of Cu_NFe (N = 1–5) Nanoalloy Clusters

PRABHAT RANJAN[1] and TANMOY CHAKRABORTY[2*]

[1]*Department of Mechatronics Engineering, Manipal University Jaipur, Dehmi Kalan, Jaipur-303007, India*

[2]*Department of Chemistry, Manipal University Jaipur, Dehmi Kalan, Jaipur-303007, India*

Corresponding author. E-mail: tanmoychem@gmail.com; tanmoy.chakraborty@jaipur.manipal.edu

ABSTRACT

Nowadays, the study based on bimetallic nanoalloy clusters have received a lot of attention due to its wide range of applications in science and technology. Addition of single-impurity atom in the pure cluster enhances the stability and physicochemical properties of pure cluster. In this report, Fe-doped Cu clusters have been studied by using density functional theory (DFT) methodology. The DFT-based global descriptors; namely, HOMO–LUMO energy gap, hardness, softness, electronegativity, electrophilicty index and dipole moment of Cu_nFe (n = 1–5) clusters have been computed. The result reveals that HOMO–LUMO energy gap has direct relationship with molecular hardness and inverse relationship with softness values. The linear correlation between DFT-based descriptors and HOMO–LUMO energy gap supports our study.

12.1 INTRODUCTION

In recent decade, the bimetallic nanoalloy clusters have gained a lot of popularity due to its large range of technological applications.[1–5] It has been reported that doping of single-impurity atom may enhance the structural, electronic, optical, catalytic, and magnetic properties of pure clusters.[4–6] Noble-metal clusters can be broadly applied for technological applications in nanoscience, bio-physics, microelectronics, and material sciences due to their remarkable electronic, optical, and magnetic properties.[7–13] The progressive conjoint effects of two or more noble metals on these above-mentioned properties have been clearly explicated by the investigators.[8,14–16] In recent years, different compositions and conformations of nanoalloys are being developed for advancement of methodologies and characterization techniques.[8,14,16] The study associated with core-shell structure of nano clusters is becoming popular because their physical and chemical properties can be improved by proper control of other structural and chemical parameters. In the periodic table, Group 11 (Cu, Ag, and Au) metallic clusters display the filled inner d orbitals with single unpaired electron in the valence s shell.[17] This procedure is responsible for the reproduction of precisely equivalent shell effects,[18–22] which are experimentally observed for alkali metal clusters.[23–25] Zhang et al.[26] have studied pure copper doped with Fe atom due to its interesting magnetic properties. The result reveals that Kondo effect decreased to 0.3% in iron doped Cu clusters and this process has shown reduction of the Kondo screening cloud, which should be lesser than the particle size.[27] In the case of $Cu_{54}Fe$ cluster, 3d magnetic moments of copper atom are in parallel with the 3d magnetic moments of Fe atom, whereas the 4p magnetic moments of copper and iron atom are in antiparallel.[28] The $Cu_{12}Fe$ clusters have demonstrated different local magnetism properties from the pure copper clusters and Al12 clusters.[29] Ling et al.[30] have reported the structural, electronic, and magnetic properties of Fe doped Cu Cu_nFe (n =1–12) clusters by using Density Functional Theory methodology. The clusters exhibit planar geometry for n = 2–5 and 3D geometry for n = 6–12. The structural properties show that iron atom is likely to occupy the location with the high coordination number. It is well known fact that due to size and surface effects, small clusters have unique physicochemical properties, which are entirely different from their bulk counterparts. Recently, we have also studied the structural, electronic, and optical properties of bimetallic nanoalloy clusters based on DFT method in which it has been observed that impurity atom enhances the stability and physicochemical properties of host clusters.[31–41] In this report, we have systematically studied small iron doped copper Cu_nFe (n =1–5) nanoalloy clusters invoking DFT-based descriptors.

12.2 COMPUTATIONAL METHODOLOGY

In this report, a computational analysis has been performed on Au$_n$Pt (n = 1–8) nanoalloy clusters invoking electronic structure theory. The geometry optimizations of all the clusters have been done using Gaussian 03[42] within DFT framework. For computation, Local Spin Density Approximation (LSDA) exchange correlation with basis set LanL2dz[43] has been chosen.

The ionization energy (*I*) and electron affinity (*A*) of all the nanoalloy clusters have been calculated by using Koopman's approximation in terms of following ansatz[44]:

$$I = -\varepsilon_{HOMO} \tag{12.1}$$

$$A = -\varepsilon_{LUMO} \tag{12.2}$$

By using, *I* and *A*, DFT-based global descriptors; namely, electronegativity (χ), molecular hardness (η), softness (*S*), and electrophilicity index (ω) have been calculated, which are as follows:

$$\chi = -\mu = \frac{I + A}{2} \tag{12.3}$$

Where, μ is the chemical potential of the system.

$$\eta = \frac{I - A}{2} \tag{12.4}$$

$$S = \frac{1}{2 * \eta} \tag{12.5}$$

$$\omega = \frac{\mu^2}{2 * \eta} \tag{12.6}$$

12.3 RESULTS AND DISCUSSION

A theoretical analysis of Cu$_n$Fe (n = 1–5) nanoalloy clusters has been investigated by using DFT methodology. The orbital energies in form of Highest Occupied Molecular Orbital (HOMO)–Lowest Unoccupied Molecular Orbital (LUMO) energy gap along with computed DFT-based global descriptors; namely, molecular hardness, softness, electronegativity, electrophilicity index and dipole moment of Cu$_n$Fe (n = 1–5) clusters have been calculated. The HOMO–LUMO energy gap along with computed DFT-based descriptors are presented in Table 12.1.

TABLE 12.1 Computed DFT-Based Descriptors along with HOMO–LUMO Energy Gaps of Cu_nFe (n = 1–5) Nanoalloy Clusters.

Species	HOMO–LUMO gap (eV)	Electrone-gativity (eV)	Hardness (eV)	Softness (eV)	Electro-philicity index (eV)	Dipole moment (Debye)
CuFe	0.844	4.476	0.422	1.186	23.752	0.717
Cu_2Fe	0.571	5.075	0.286	1.750	45.068	0.817
Cu_3Fe	1.061	3.633	0.531	0.942	12.434	0.391
Cu_4Fe	0.990	5.034	0.495	1.010	25.584	0.783
Cu_5Fe	0.327	4.925	0.163	3.063	74.286	0.375

The HOMO–LUMO energy gap is an essential parameter to compute the electronic properties of nanoalloy clusters. It specifies the energy required for an electron to jump from occupied orbital to unoccupied orbital. The HOMO–LUMO energy gaps have noticeable impact on chemical stability of the clusters. The high value of HOMO–LUMO energy gap indicates the stability of clusters whereas, lowest value of energy gap specifies the maximum response to an external perturbation. The result obtained from Table 12.1 shows that cluster Cu_3Fe has maximum HOMO–LUMO energy gap, whereas cluster Cu_5Fe exhibits minimum energy gap in this range of molecular system. The result also describes that HOMO–LUMO energy gaps of Cu_nFe nanoalloy clusters have linear relationship with molecular hardness and inverse relationship with softness values. It indicates that HOMO–LUMO energy gap and hardness value runs hand in hand. This is a probable propensity from experimental point of view also, as the frontier orbital energy gap increases, their molecular hardness value also increases. The similar kind of relationship is observed in case of other DFT-based descriptors along with their HOMO–LUMO energy gap. The linear correlation between DFT-based descriptors and HOMO–LUMO energy gap are shown in the Figures 12.1 and 12.2. The correlation coefficient $R^2 = 0.912$ is obtained between Softness and HOMO–LUMO energy gap from Figure 12.1 whereas, $R^2 = 0.940$ is achieved between electrophilicity index and HOMO–LUMO energy gap from Figure 12.2. The maximum value ($R^2 = 1$) of correlation coefficient is obtained between hardness and HOMO–LUMO energy gap whereas, minimum value ($R^2 = 0.3$) obtained between dipole moment and HOMO–LUMO energy gap.

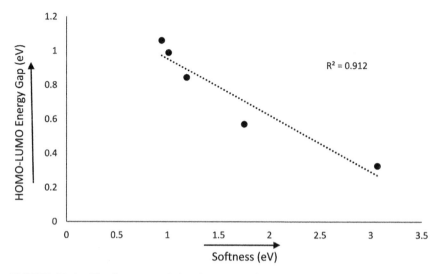

FIGURE 12.1   The linear correlation between softness versus HOMO–LUMO energy gap of Cu$_n$Fe nanoalloy clusters (n = 1–5).

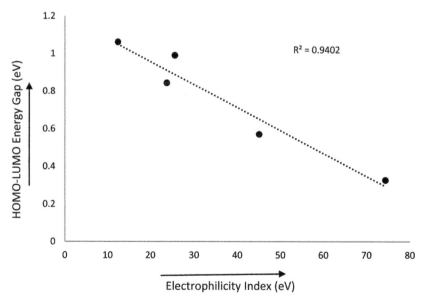

FIGURE 12.2   The linear correlation between electrophilicity index versus HOMO–LUMO energy gap of Cu$_n$Fe nanoalloy clusters (n = 1–5).

12.4   CONCLUSION

In this report, a theoretical study has been performed on Fe-doped Cu nano-alloy clusters Cu_nFe (n = 1–5) by using DFT methodology. The DFT-based global descriptors; namely, molecular hardness, softness, electronegativity, electrophilicity index and dipole moment have been calculated along with their HOMO–LUMO energy gap. The result indicates that clusters Cu_3Fe have maximum HOMO–LUMO energy gap, whereas cluster Cu_2Fe shows minimum energy gap in this range. The computed result implies that HOMO–LUMO energy gap runs hand in hand along with the molecular hardness of Cu_nFe clusters. This is a probable propensity from experimental point of view also, as the frontier orbital energy gap increases, their molecular hardness value also increases. The computed regression coefficient between DFT-based descriptors and HOMO–LUMO energy gap supports our study.

KEYWORDS

- **DFT**
- **bimetallic clusters**
- **Cu-Fe**
- **hardness**
- **softness**
- **HOMO–LUMO energy gap**

REFERENCES

1. Heer, W. A. The Physics of Simple Metal Clusters: Experimental Aspects and Simple Models. *Rev. Mod. Phy.* **1993,** *65* (3), 611–676.
2. Bonačiæ-Koutecký, V.; Fantucci, P.; Koutecký, J. Quantum Chemistry of Small Clusters of Elements of Groups Ia, Ib, and IIq: Fundamentals Concepts, Predictions, and Interpretation of Experiments. *Chem. Rev.* **1991,** *91* (50), 1035–1108.
3. Morse, M. D. Clusters of Transition-Metal Atoms. *Chem. Rev.* **1986,** *86* (6), 1049–1109.
4. Alonso, J. A. Electronic and Atomic Structure, and Magnetism of Transition-Metal Clusters. *Chem. Rev.* **2000,** *100* (2), 637–678.
5. Li, X.; Kuznetsov, A. E.; Zhang, H. F.; Boldyrev, A. I.; Wang, L. S. Observation of All-Metal Aromatic Molecules. *Science* **2001,** *291* (5505), 859–861.

6. Janssens, E.; Neukermans, S.; Lievens, P. Shell of Electrons in Metal Doped Simple Metal Clusters. *Curr. Opin. Solid State Mater. Sci.* **2004**, *8* (3), 85–193.
7. Teng, X.; Wang, Q.; Liu, P.; Han, W.; Frenkel, A. I.; Wen, M. N.; Hanson, J. C.; Rodriguez, J. A. *J. Am. Chem. Soc.* **2008**, *130*, 1093.
8. Ferrando, R.; Jellinek, J.; Johnston, R. L. *Chem. Rev.* **2008**, *108*, 845.
9. Henglein, A. *J. Phys. Chem.* **1993**, *97*, 5457.
10. Davis, S. C.; Klabunde, K. J. *Chem. Rev.* **1982**, *82*, 153.
11. Lewis, L. N. *Chem. Rev.* **1993**, *93*, 2693.
12. Schmid, G. *Chem. Rev.* **1992**, *92*, 1709.
13. Schon, G.; Simon, U. Colloid. *Polym. Sci.* 1995, *273*, 101.
14. Oderji, H. Y.; Ding, H. *Chem. Phys.* **2011**, *388*, 23.
15. Liu, H. B.; Pal, U.; Medina, A.; Maldonado, C. *J. A. Ascencio, Phys. Rev.* **2005**, *B 71*, 075403.
16. Baletto, F.; Ferrando, R. *Rev. Mod. Phys.* **2005**, *77*, 371.
17. Alonso, J. A. *Chem. Rev.* **2000**, *100*, 637.
18. Katakuse, I.; Ichihara, T.; Fujita, Y.; Matsuo, T.; Sakurai, T.; Matsuda, H. *Int. J. Mass Spectrom. Ion Processes* **1985**, *67*, 229.
19. Katakuse, I.; Ichihara, T.; Fujita, Y.; Matsuo, T.; Sakurai, T.; Matsuda, H. *Int. J. Mass Spectrom. Ion Processes.* **1986**, *74*, 33.
20. Heer, W. A. D. *Rev. Mod. Phys.* **1993**, *65*, 611.
21. Gantefor, G.; Gausa, M.; Meiwes-Broer K. H.; Lutz, H. O. *J. Chem. Soc. Faraday Trans.* **1990**, *86*, 2483.
22. Leopold, D. G.; Ho, J.; Lineberger, W. C. *J. Chem. Phys.* **1987**, *86*, 1715.
23. Lattes, A.; Rico, I.; Savignac, A. D.; Samii, A. A. Z. *Tetrahedron* **1987**, *43*, 1725.
24. Chen, F.; Xu, G. Q.; T. S. A. *Hor. Mater. Lett.* **2003**, *57*, 3282.
25. Taleb, A.; Petit, C.; Pileni, M. P. *J. Phys. Chem.* **1998**, *B 102*, 2214.
26. Zhang, R.; George, T. A.; Kharel, P.; Skomski, R.; Sellmyer, D. J. Susceptibility of Fe Atoms in Cu Clusters. *J. Appl. Phys.* **2013**, *113*, 17E148.
27. Kang, B. S.; Lee, H. K.; Sohn, K. S. Spin Polarization in Dilute Cu-Fe Clusters. *J. Korean Phys. Soc.* **1997**, *31*, 660–663.
28. Sun, Q.; Gong, X. G.; Zheng, Q. Q.; Sun, D. Y.; Wang, G. H. Local Magnetic Properties and Electronic Structures of 3d and 4d Impurities in Cu Clusters. *Phys. Rev.* **1996**, *B*, *54*, 10896–10904.
29. Khuntia, P.; Mariani, M.; Mozzati, M. C.; Sorace, L.; Orsini, F.; Lascialfari, A.; Borsa, F.; Maxim, C.; Andruh, M. Magnetic Properties and Spin Dynamics in the Single-Molecule Paramagnets Cu6Fe and Cu6CO. *Phys. Rev.* **2009**, *B*, *80*, 094413.
30. Ling, W.; Dong, D.; Shi-Jian, W.; Zheng-Quan, Z. Geometrical, Electronic and Magnetic Properties of CunFe (n=1–12) Clusters: A Density Functional Study. *J. Phy. Chem. Solids*, **2015**, *76*, 10–16.
31. Ranjan, P.; Kumar, A.; Chakraborty, T. Computational Study of Nanomaterials Invoking DFT-Based Descriptors. In *Environmental Sustainability: Concepts, Principles, Evidences and Innovations*; G. M. Mishra Ed.; Excellent Publishing House: New Delhi, 2014; pp. 239–242.
32. Ranjan, P.; Dhail, S.; Venigalla, S.; Kumar, A.; Ledwani, L.; Chakraborty, T. A Theoretical Analysis of Bi-Metallic (Cu-Ag) n=1–7 Nano Alloy Clusters Invoking DFT-Based Descriptors. *Mater. Sci-Poland* **2015**, *33* (4), 719–724.

33. Ranjan, P.; Kumar, A.; Chakraborty, T. Computational Study of AuSin (n=1–9) Nanoalloy Clusters Invoking DFT-Based Descriptors. *AIP Conference Proceedings* **2016**, *1724*, 020072.

34. Ranjan, P.; Kumar, A.; Chakraborty, T. Theoretical Analysis: Electronic and Optical Properties of Gold-Silicon Nanoalloy Clusters. *Mater. Today Proceedings* **2016**, *3*, 1563–1568.

35. Ranjan, P.; Chakraborty, T.; Kumar, A. Computational Investigation of Cationic, Anionic, and Neutral Ag2AuN (N=1–7) Nanoalloy Clusters. *Phys. Sci. Rev.* **2017**, *2* (10), 20160112.

36. Ranjan, P.; Venigalla, S.; Kumar, A.; Chakraborty, A. Theoretical Study of Bi-Metallic Ag_mAu_n (m+n=2–8) Nano Alloy Clusters in Terms of DFT-Based Descriptors. *Front. Chem.* **2014**, *23* (2), 111–122.

37. Ranjan, P.; Chakraborty, T. A DFT Study of Vanadium Doped Gold Nanoalloy Clusters. *Key Eng. Mater.* **2018**, *777*, 183–189.

38. Ranjan, P.; Kumar, A.; Chakraborty, T. Computational Study of AumSin (m+n=2–6) Nanoalloy Clusters Invoking Density Functional Based Descriptors. *J. Phy.: Conference Series* **2016**, *759*, 012045.

39. Ranjan, P.; Kumar, A.; Chakraborty, T. Computational Investigation of Ge doped Au Nanoalloy Clusters: A DFT Study. *Mater. Sci. Eng.: IOP Conference Series* **2016**, *149*, 012172.

40. Ranjan, P.; Chakraborty, T. Density Functional Approach: To Study Copper Sulfide Nanoalloy Clusters. *Acta Chim. Slov.* **2019**, *66*, 173–181.

41. Ranjan, P.; Chakraborty, T.; Kumar, A. Computational Study of Au Doped Cu Nano Alloy Clusters. *Nano Hybrids and Composites* **2017**, *17*, 62–71.

42. Gaussian 03, Revision C.02, Frisch, M. J.; Trucks, G. W.; Schlegel, H. B.; Scuseria, G. E.; Robb, M. A.; Cheeseman, J. R.; Montgomery, Jr., J. A.; Vreven, T.; Kudin, K. N.; Burant, J. C.; Millam, J. M.; Iyengar, S. S.; Tomasi, J.; Barone, V.; Mennucci, B.; Cossi, M.; Scalmani, G.; Rega, N.; Petersson, G. A.; Nakatsuji, H.; Hada, M.; Ehara, M.; Toyota, K.; Fukuda, R.; Hasegawa, J.; Ishida, M.; Nakajima, T.; Honda, Y.; Kitao, O.; Nakai, H.; Klene, M.; Li, X.; Knox, J. E.; Hratchian, H. P.; Cross, J. B.; Bakken, V.; Adamo, C.; Jaramillo, J.; Gomperts, R.; Stratmann, R. E.; Yazyev, O.; Austin, A. J.; Cammi, R.; Pomelli, C.; Ochterski, J. W.; Ayala, P. Y.; Morokuma, K.; Voth, G. A.; Salvador, P.; Dannenberg, J. J.; Zakrzewski, V. G.; Dapprich, S.; Daniels, A. D.; Strain, M. C.; Farkas, O.; Malick, D. K.; Rabuck, A. D.; Raghavachari, K.; Foresman, J. B.; Ortiz, J. V.; Cui, Q.; Baboul, A. G.; Clifford, S.; Cioslowski, J.; Stefanov, B. B.; Liu, G.; Liashenko, A.; Piskorz, P.; Komaromi, I.; Martin, R. L.; Fox, D. J.; Keith, T.; Al-Laham, M. A.; Peng, C. Y.; Nanayakkara, A.; Challacombe, M.; Gill, P. M. W.; Johnson, B.; Chen, W.; Wong, M. W.; Gonzalez, C.; and Pople, J. A.; Gaussian, Inc.: Wallingford CT, 2004.

43. Wang, H.-Q.; Kuang, X.-Y.; Li, H.-F. Density Functional Study of Structural and Electronic Properties of Bimetallic Copper-Gold Clusters: Comparison with Pure and Doped Gold Clusters. *Physical Chem. Chem. Phy.* **2010**, *12*, 5156–5165.

44. Parr, R. G.; Yang, W. *Density Functional Theory of Atoms and Molecules*, Oxford University Press: New York, Oxford, 1989.

CHAPTER 13

Nanotechnology as a Clean Technology and a Vision for the Future

SUKANCHAN PALIT[1] and CHAUDHERY MUSTANSAR HUSSAIN[2*]

[1]*Department of Chemical Engineering, University of Petroleum and Energy Studies, Energy Acres, Dehradun 248007, Uttarakhand, India*

[2]*Department of Chemistry and Environmental Sciences, New Jersey Institute of Technology, University Heights, Newark, NJ 07102, USA*

*Corresponding author. E-mail: chaudhery.m.hussain@njit.edu

ABSTRACT

Scientific progress in the field of nanotechnology is in the path of greater emancipation. Nanomaterials and engineered nanomaterials applications are today considered to be clean technology. In this paper, the authors deeply investigate the world of green nanotechnology, green sustainability, and environmental protection science. The authors also touched upon environmental biotechnology and its immense vision. Scientific research pursuit in nanomaterials will surely usher in a new epoch as civilization moves forward. There is a definite vision in the application of nanotechnology in human society today. The environmental ethics and health issues concerning application of nanomaterials needs to be addressed with each step of human scientific progress. The other areas of scientific pursuit are environmental protection and the vast scientific interface of environmental engineering with nanotechnology. Today, the world of science and engineering are in the avenues of newer scientific grit, determination, and profundity. This chapter will surely go a long way in uncovering the scientific truth behind nanotechnology applications in diverse area of science and engineering.

13.1   INTRODUCTION

Nanotechnology today in the global scenario is in the forefront of a scientific revolution. In the similar vein, nanomaterials, eco-materials, and green materials are the veritable needs of civilization today. Global climate change, frequent environmental disasters, and depletion of fossil fuel resources are really challenging the vast and versatile scientific firmament. Today, there is a serious debate as to whether nanomaterials is a clean technology. Nanomaterials, engineered nanomaterials, and smart materials are in the forefront of scientific discussion today. In this paper, the authors have deeply discussed with scientific vision and scientific provenance the vision of clean nanotechnology in the true furtherance of science and engineering globally. Clean technology, environmental engineering tools, and built environment are the coinwords of global environmental concerns in the present time. Nanotechnology and nanomaterials are creating scientific wonders. This paper discussed with vision and insight the challenges, the scientific stance, and the scientific ingenuity in the true emancipation of environmental sustainability also. Developing and industrialized nations around the world are at the forefront of a major environmental disaster, that is, arsenic and heavy metal groundwater contamination. This is a monstrous issue which has practically no solution. Here comes the importance and vision of nanoscience and nanotechnology. The scientific grit and integrity in research and development of clean technologies and environmental protection will all be the pallbearers toward newer thoughts and new future recommendations in groundwater decontamination. The applications of nanomaterials and engineered nanomaterials in water and wastewater treatment are the wonders and the zenith of global research and development initiatives in water science and technology. The ethical and the health and safety issues of nanomaterial applications to human society is an important area of scientific endeavor today. The challenges, the scientific divination, and revelation are vastly revealed as the authors trudge a difficult path in the field of nanoscience and nanotechnology. This paper will surely open up new doors of scientific instinct and profundity in the field of nanotechnology and nano-engineering in years to come. The scientific challenges, the scientific ardor, and the future recommendations are discussed in minute details in this paper.[27,28]

13.2 THE AIM AND OBJECTIVE OF THIS PAPER

Civilization today is surpassing one scientific boundary over another. Scientific acuity and vast scientific articulation are the utmost needs of research and development initiatives globally today. The main aim and objective of this study is to glean and depict with cogent insight the needs of nanomaterials and engineered nanomaterials in the success of science and engineering globally. The authors describe and delineate the recent developments in the field of nanomaterials, nanotechnology, and environmental remediation. A new dawn in the field of environmental protection will surely usher in with immense vision and might as civilization treads forward. Arsenic and heavy metal groundwater contamination are devastating the global scientific fabric. This paper will veritably open up new discourses, new thoughts, and future recommendations in the field of environmental and chemical engineering tools. The scientific truth in the ethical and environmental issues of nanotechnology applications are revealed and deeply investigated in this paper. The authors do not leave any concepts uncovered in the field of conventional and nonconventional environmental engineering techniques. The vision of this paper is to investigate different types of nanomaterials and engineered nanomaterials in chemical engineering and environmental engineering processes. Adsorption is a removal technique of hazardous materials in water and wastewater. These areas are depicted in details in this paper. Environmental nanotechnology is a wide branch of science and technology. Scientific introspection and deep scientific enquiry in the field of environmental nanotechnology, nanobiotechnology, and biotechnology are dealt in details in this paper.[27,28]

13.3 CLEAN NANOMATERIALS

Clean technology and green sustainability are the needs of civilization and scientific progress today. Nanomaterials and engineered nanomaterials are today in the forefront of a new scientific revolution. The application areas of nanomaterials and nanotechnology today needs to be re-envisioned and reframed as civilization trudges forward toward a new visionary era. The ethical and health aspects of nanomaterials and engineered nanomaterials are still latent and needs to be unraveled. Technological and scientific validation and the futuristic vision of nanotechnology will veritably open up new thoughts and newer future recommendations in research in the field of nanomaterials in decades to come. Today, there is an ongoing debate as to whether nanotechnology is a clean technology. Scientists and engineers are at the forefront of a global debate on the ethics of application of nanotechnology in human society. Nanomaterials and engineered nanomaterials have tremendous applications in every branch of science and engineering such as environmental protection and health sciences. In this paper, the author veritably reiterates the success, the ingenuity, and the scientific integrity in the application of nanomaterials in the progress of civilization. Clean nanomaterials are really creating wonders in scientific progress of environmental engineering, chemical engineering, and medical science. So, the debate is justified as to whether nanotechnology is clean. Here comes the importance and effort of environmental engineers and medical practitioners. The concepts of environmental remediation and chemical process engineering need to be re-envisioned with the passage of history and time.

The Royal Society, United Kingdom Report (2004)[1] discussed with immense scientific conscience and far-sightedness about nanoscience and nanotechnologies and their opportunities and uncertainties. This report delineates the vast domain of nanoscience and nanotechnologies, science and applications, nanomanufacturing, and the industrial application of nanotechnologies, adverse health, environmental and safety impacts, social and ethical issues, stakeholder and public dialogue, and finally the regulatory issues.[1] Nanoscience and nanotechnologies are widely seen as having immense potential to bring benefits to many areas of scientific research and applications, and are attracting rapidly increasing investments from governments around the world.[1,27,28] Technological verve and drive, the vast domain of scientific validation, and the futuristic vision of nanotechnology will surely open up new doors of scientific innovation in decades to come.[1] The remit and purpose of this study is:

- Define what is meant by nanoscience and nanotechnologies.
- Summarize the current state of knowledge in nanotechnology.
- Identify the specific applications of nanotechnologies.[1]
- Carry out a forward look to see how the technologies might be used in future.
- Identify health and safety, environmental, and ethical implications of nanotechnology applications.
- Identify areas where additional regulations need to be considered.[1]

Civilization today stands in the crucial juxtaposition of forbearance, might, and vision. The world of science and technology of nanomaterials applications in human society needs to be reorganized as civilization moves forward.

13.4 ENVIRONMENTAL SUSTAINABILITY AND THE VAST VISION FOR THE FUTURE

Sustainability, whether it is energy or environmental, is the necessity of civilization today. Mankind's vast scientific prowess and civilization's vast knowledge dimensions as regards sustainable development needs to be reframed with the march of science. Human challenges and scientific divination are two opposite sides of the coin in the global scenario. Environmental sustainability in poor and developing nations around the world are in a latent stage and needs to be re-envisioned as science and technology moves forward. In the similar vein, environmental sustainability should be applied and envisioned in every stage of human scientific progress. Social and economic sustainability are the other sides of the coin. Today, environmental remediation science and industrial pollution control are in the middle of vast scientific introspection and deep comprehension. The visionary words of Dr. Gro Harlem Brundtland, former Prime Minister of Norway on the science of "sustainability" needs to be restructured and reframed as civilization trudges forward.

United Nations Development Programme Report (2011)[2] depicts and investigates profoundly sustainability and inequality in human development. This paper analyzes the theoretical and empirical links between inequality in human development on the one side of the visionary coin and sustainability on the other hand. Technology and engineering have advanced so rapidly in present day human civilization that sustainability is the need of the hour.[2] It specifically looks at the causes in both the research directions. Inequality in various dimensions of human development are analyzed with deep scientific conscience along with weak and strong sustainability where weak sustainability presumes substitutability among different forms of capital while strong sustainability rejects substitutability and envisions preservation of so-called critical forms of natural capital.[2] Many academics and intellectuals have in recent times voiced grave concerns about increasing inequality and its detrimental social effects. It is slowly decreasing the "strong fabric" and the public spirit on the platform on which private and public welfare of all societies are strongly built.[2] This paper analyzes the links between inequality in human development on the one hand and sustainable development on the other hand. Enabling everyone to be capable and free to do things and be the person they want to be is the holistic goal of human development.[2] Technological and scientific challenges are thus the platforms of sustainability today. If the progress of science needs to be envisioned, then sustainability is the immediate need, whether it is energy or environmental sustainability. The authors deeply investigate weak and strong sustainability and comprehend the link between weak, strong sustainability, and the holistic world of environmental protection. Human civilization, today, is befallen as scientific difficulties surmounts in the application of energy and environmental sustainability to human society. Thus, there is a need of a well-researched treatise.

Markulev et al. (2013)[3] discusses with scientific far-sightedness the economic approach of sustainability. Sustainability is envisioned as a desirable objective in a wide range of contexts, yet its meaning is not clear. At its most general level, sustainability refers to the capacity to continue an activity or process indefinitely.[3] The most frequently cited definition is that of the United Nations World Commission on Environment and Development (1987; the 'Brundtland' Commission)—Development that meets the needs of the present without compromising the ability of future generations to meet their own needs.[3] In Australia, the National Strategy for Ecological Sustainable Development defines sustainable development as: "using , conserving and enhancing the community's resources so that ecological processes are well maintained and the total quality of life, now and future, can be increased." As a developed nation, the sustainability strategy of Australia is an example for other developing and developed countries around the world. The authors of this report discuss sustainability from an economic perspective.[3] The concept of wellbeing is a veritable need in defining sustainability. The concept of "sustainability" in Australian Government policies is the other hallmark of this report. Economic and social sustainability in global context are the other cornerstones of this well-researched report.[3]

Kates et al. [4] discussed with cogent insight about sustainable development and its goals, indicators, values, and practices. Humanity has the ability and scientific might to make development sustainable—to ensure that it meets the needs of the present without compromising the ability of future generations to meet their own needs. Environmental sustainability and environmental protection are two opposite sides of the visionary coin.[4] The World Commission on Environment and Development was initiated by the General Assembly of the United Nations in 1982 and its report, "Our Common Future," was published in 1987. It was chaired by then-Prime Minister of Norway Gro Harlem Brundtland, thus envisioning the name "The Brundtland Commission."[4] According to the report, over a period of time, peace, freedom, development, and aspiration remain prominent aspirations and definite issues. In the years following Brundtland Commission Report, water-shed texts in the field of sustainable development came out after lots of scientific and technological deliberations.[4] The United States National Academy of Sciences Report brought forward to the scientific panorama a report titled "Our Common Journey: A Transition Toward Sustainability." The study of the eco-system services stands as a major pillar in the journey toward sustainable development. Civilization's travails, the futuristic vision of science and engineering, and the world of sustainable development will surely unravel the difficulties of environmental and energy sustainability. The destruction of ecosystem globally is a bane of mankind today. The United Nations Millennium Declaration in September, 2000 unraveled the scientific intricacies and the scientific barriers of Global Sustainable Development Goals.[4] Engineering and technological vision needs to be envisioned if sustainability is to be maintained in the global scenario. According to the report, sustainability is a social movement in the global scenario. It is an integrated effort by a group of people serving the human society.[4] The global solidarity movement encompasses support for poor people in developing nations around the world that goes beyond development funding. Here comes also the need of science and technology in true emancipation of sustainable development.[4]

Kuhlman et al. (2010)[5] discussed with vast scientific vision the domain of sustainability. Sustainability as a policy concept has its origin in the Brundtland Report of 1987. Technological vision and verve, the futuristic vision of sustainability, and the needs of human society will all open up new dimensions in the field of environmental and energy sustainability.[5] The Brundtland Report was concerned with the tension between the aspirations of mankind toward a better life on one hand and the limitations of nature on the other hand. Civilization is in deep peril as global climate change and loss of ecological biodiversity haunts the vast scientific firmament. Thus, the need of sound scientific vision and deep scientific integrity.[5] This sustainability concept has been re-interpreted as encompassing three knowledge dimensions, namely, social, economic, and environmental. The debate on sustainability has been cast into a distinction between "weak" and "strong" sustainability. The term "sustainability" has become popular in policy-oriented research. The principal inspiration came from Brundtland Commission Report. Civilization's immense scientific integrity and profundity will lead toward a new era in the field of sustainable development globally.[5] The authors targeted the concept of people, planet, and profit in the true scientific understanding of sustainability.[5]

13.5 APPLICATION OF NANOTECHNOLOGY AND CLEAN TECHNOLOGY IN WATER AND WASTEWATER TREATMENT

Human civilization and its scientific and academic rigor are in the middle of immense scientific provenance and deep divination. Water and wastewater treatment are in the similar vein in the midst of vast scientific introspection and immense vision and acuity. Engineering prowess and scientific discernment are the true needs of research and development endeavor today. In this section, the authors discuss some of the major roles played by nanotechnology in greater emancipation of environmental remediation science.

Werkneh et al. (2019)[6] discussed with vast scientific vision, alacrity, and foresight the applications of nanotechnology and biotechnology for sustainable water and wastewater treatment. The vast knowledge prowess in the field of environmental remediation is the need of the hour. Nowadays, water pollution and freshwater scarcity have become a serious problem worldwide causing concern to human health and public health engineering.[6] The vision and provenance of science and engineering are the pallbearers toward a new era in environmental protection today.[6] To reduce the environmental challenges, various treatment technologies have been envisioned. Among these technologies, nanotechnology and biotechnology-based tools are usually used separately for water (domestic) purposes and wastewater (reuse) treatment. Civilization's scientific fervor and deep ardor are the necessities of science and engineering today.[6] This chapter focuses on new and emerging nano- and biotechnologies for the sustainable removal of hazardous pollutants during water and wastewater treatment. Besides the toxicological and safety aspects of different nanotechnologies and their current and vast future perspectives are discussed in minute details.[6] Environmental pollution together with global warming is one of the foremost challenges of the

21st century. The 21st century is an age of immense scientific and engineering wonders. Environmental engineering and environmental protection are also in the path of new wonders. Besides access to fresh water in many developing and disadvantaged countries around the world is affecting millions of families worldwide and it is still a threatening issue that needs to be solved with scientific might and ingenuity. The vast and extensive industrial development, rapid population growth, and global climate change contributes largely to the deterioration of the physiochemical and biological characteristics of the available water resources.[6] From the industrial sectors and due to the lack of improved water supply and sanitation systems, high quantities of hazardous pollutants are discharged to the surrounding environment every day.[6] The conventional water-treatment technologies used for the remediation of water pollutants are the activated carbon-based adsorption, membrane filtration, ion exchange, coagulation and flocculation, reverse osmosis, flotation and extraction, electrochemical treatment, advanced oxidation processes, and biosorption that are widely used in several environmental research situations.[6] Environmental engineering research endeavor are today in the avenues of deep scientific enquiry and scientific prowess. Most of the conventional environmental engineering techniques have serious drawbacks in terms of its operational methods, energy requirements, processing efficiency, and economic benefits thereby further reducing large-scale and long-term applicability. Thus, the need of nonconventional environmental engineering tools has emerged.[6] From an environmental sustainability point of view, the use of microbes and nanomaterials for the removal of hazardous pollutants has received tremendous attention from environmental engineers and scientists. Nanomaterials are very small in size, that is, approximately 1–100 nm and shows splendid and unique characteristics that enables them to use in wide range of scientific applications such as in wastewater treatment and others. They exhibit high surface area-to-volume ratio, which is very significant to produce high reactive surface area than the bulk counterparts. The validation of the science of nanotechnology is the need of human scientific progress and march of civilization.[6] Nano-oxides (silver, gold, iron, and titanium) are extremely common nanomaterials, which have been used for the remediation of hazardous pollutants in contaminated water and soil environments.[6] The targets and the challenges of environmental remediation are immense and versatile today.[6] The authors in this chapter also touched upon environmental biotechnology, which offers to provide human society feasible scientific solutions and answers. The authors also dealt with sustainable water and wastewater treatment technologies, the nanotechnology perspectives, nanomaterials for the disinfection of pathogenic microbes, photocatalytic applications of nanomaterials, applications of nanomaterials as adsorbents, the environmental biotechnology perspectives, bioremediation, and biotransformation techniques.[6] Bioreactor configurations in water and wastewater treatment are the other pillars of this scientific treatise. A case study in water and wastewater treatment and the toxicological perspectives on nano/biotechnology stands in science with immense scientific introspection. Nanotechnology and biotechnology are two highly promising technologies in 21st century science and engineering, especially in the field of water and wastewater treatment. Nanotechnologies have shown immense scientific and engineering prowess and have demonstrated higher removal efficiency of the pollutants from water and wastewater. Due to their nanoscale size, the veritable assessment and management of the associated risk are often challenging and highly limited.[6] The surge in science and technology of nano-engineering will inevitably open up new doors of innovation and scientific instinct in the world of biotechnology and environmental biotechnology in decades to come. This chapter is an integration of the science and engineering of environmental biotechnology. Further research recommendations of this study include: (1) the potential hazards of these materials in water and wastewater treatment, (2) to integrate innovative nanotechnology and biotechnology, and (3) long-term performance of biotechnology techniques. The application of chemical agents such as nanomaterials and biocatalysts in an integrated approach are the other hallmarks of this chapter.[6] The world of scientific challenges and the knowledge prowess in the field of biotechnology will surely widen human scientific thoughts in nanotechnology and biotechnology.

Kunduru et al. (2017)[7] deeply discussed with vision and scientific far-sightedness the nanotechnology for water purification and application of nanotechnology methods in wastewater treatment. Water is the most important asset of human mankind and is the utmost necessity of human life. At present the demand is higher than the supply of drinking water.[7] Water and wastewater treatment today is in the midst of deep scientific contemplation and vision. Demand for water is escalating due to population growth, global warming, and water-quality deterioration. Only 2.5% of the world's oceans and seas have fresh water (salts concentration less than 1 g/L). However, 70% of fresh water is found to be frozen as eternal ice.

Only less than 1% of fresh water can be used for drinking. Globally, greater than 700 million people do not have access to potable water.[7] This is a problem in developing countries and sub-Saharan African countries. Water contaminants may be organic, inorganic, and biological. Some contaminants can be highly toxic and carcinogenic and have dangerous effects on human beings and ecosystems. Technological and scientific impasse are at its zenith as human scientific progress treads forward toward a new visionary era.[7] Arsenic is one of the deadliest element well known since ancient times. Technology and engineering should be reframed and reorganized as regards groundwater remediation.[7] Other heavy metal water pollutants with high toxicity are cadmium, chromium, mercury, lead, zinc, nickel, copper, and so on. Nitrates, sulphates, phosphates, fluorides, chlorides, selenides, chromates, and oxalates show severe hazardous effects at high concentrations. Man's immense scientific grit and determination and the vision of technology will all be the pallbearers toward a newer era in environmental protection science. The authors also discussed in details major limitations associated with conventional water purification methods, applications of nanotechnology in water and wastewater treatment, and a brief overview of different nanomaterials in water and wastewater treatment.

Das et al.[8] discussed with vast scientific conscience recent advances in nanomaterials for water protection and monitoring. Technological and scientific validation and vision are the hallmarks of research directions in environmental remediation today.[8] The efficient handling of wastewater pollutants is a must, since they are continuously degrading fresh water resources, seriously effecting the terrestrial, aquatic, and aerial flora and fauna.[8] The vision of this paper is to undertake an exhaustive examination of current research trends with a focus on nanomaterials to improve and enhance the performance of classical water-treatment technologies, for example, adsorption, catalysis, separation, and disinfection.[8] World population has increased at a rate of 80 billion per year, increasing potable water demand by 64 billion cubic metres per annum.[8] Science, technology, and engineering are thus in a state of immense difficulties and deep comprehension.[8] Engineering might and deep insight are thus the necessities of scientific progress in environmental protection today. The United Nations reported that almost 2 billion people do not have access to clean and safe water in 2013 and by 2025 nearly 1.8 billion people will be living under serious water scarcity.[8] Currently, more than 750 million people around the world do not have access to improved water facilities and most of them are from Asia, Central and South America, and Africa.[8]

13.5.1 *GREEN NANOTECHNOLOGY AND THE MARCH OF SCIENCE*

Green nanotechnology and the application of environmental biotechnology are the scientific prowess and scientific vision of today's research pursuit. Nanotechnology integrated with environmental engineering science will surely open up new doors of scientific enquiry in years to come. Novel separation processes, conventional, and nonconventional environmental engineering techniques are the utmost needs of science and civilization today. Thus, also comes the importance of green nanotechnology. Green nanotechnology refers to the use of nanotechnology to enhance and envision the environmental sustainability of the processes. It also refers to the use of products of nanotechnology to veritably enhance environmental sustainability. It includes making green nanoproducts and using nanoproducts in support of sustainability.[27,28] Green nanotechnology has been described as the development of clean technologies to minimize environmental risks associated with the manufacture and use of nanotechnology products, and to highly encourage replacement of existing products with more new nanoproducts that are more environmental friendly. Civilization's vast knowledge prowess, the success of science, and the world of scientific challenges will surely open up new recommendations in the field of nanoscience and nanotechnology.[27,28]

Verma et al. (2017)[9] discussed with vision and insight green nanotechnology. Nanotechnology promises a sustainable future by its growth in green chemistry to develop green nanotechnology.[9] Green nanotechnology implies the application of green chemistry and green engineering concepts in the field of nanotechnology to manipulate the molecules in a nanoscale range.[9] Technology of green nanotechnology and green chemistry is latent yet far-reaching.[9] This review investigates how nanotechnology can be highly advantageous as a green alternative in different areas of nanoparticle synthesis. Human scientific enquiry and scientific ingenuity and struggles are of a novel kind in the path toward scientific realization in green nanotechnology.[9] Green nanotechnology is based on the field of green chemistry that vastly reflects the main aim of nanotechnology to create eco-friendly nano-objects to reduce human health and environmental disasters by application of green nanoproducts. The authors of this review deeply comprehend nanoparticle synthesis by green route, challenges in green synthesis, and new approaches in green nanotechnology.[9] Nanotechnology has a revolutionary and challenging effect on many areas of science, technology, and engineering. In industry also, nanotechnology and nanomanufacturing are the pillars of scientific endeavor. The current development in the field of nanotechnology has veritably harnessed the power to convert phytochemicals into nanoparticles via green chemistry, thus ushering a new era in the field of green sustainability. The development in nanotechnology concerns in environmental engineering approach is growing at a larger extent to save fuels, reduce materials for production, monitor the trace of hazardous materials in environment and green manufacturing. Science and engineering of nanotechnology thus are in the path of new regeneration and immense scientific verve.[9] Green nanotechnology will thus open newer thoughts and newer future recommendations as civilization treads forward.[9]

Nath et al. (2014)[10] discussed with deep scientific insight green nanomaterial and how green they are as biotherapeutic tool. The vast emergence of nanoparticles has veritably attracted tremendous interest of the scientific community for decades due to their unique properties and diverse potential applications including drug delivery and therapy.[10] These opportunities are based on the unique properties such as magnetic, optical, mechanical, and electronic. Advances in nanotechnology have significantly impacted the fields of therapeutic delivery.[10] Technological and scientific ardor, vitality, and verve are the needs of research pursuit in nanotechnology today.[10] There is also a need of new discipline—nanotoxicology—that will ultimately evaluate the health effects posed by nanomaterials. Green nanotechnology gives the opportunity in lowering the risk of using nanomaterials and limiting the risk of producing nanomaterials. The authors in this review paper deeply discuss green nanotechnology and biotherapeutics. The other areas of this research endeavor are inorganic nanoparticles, and challenges of targeted drug-encapsulated nanoparticles delivery. Future opportunities of nanotherapeutic devices are the other hallmarks of this research pursuit.[10] Green chemistry metrics need to be incorporated into the vast domain of nanotechnologies at the source as health effects of nanomaterials and engineered nanomaterials assumes immense importance. This challenge and the vision is deeply investigated in this paper.[10]

Dhingra et al. (2010)[11] discussed and elucidated with scientific vision and foresight on sustainable nanotechnology through the concepts of green methods and life-cycle thinking. This paper vastly emphasizes the need to conduct life–cycle-based assessments as early in the product development process. Nanomanufacturing methods often have associated environmental and human health impacts, which must be envisioned when evaluating nanoproducts for their absolute "greenness."[11] The authors in this paper investigate nanomanufacturing methods and environmental concerns. Industrial ecology and life-cycle analysis are the other cornerstones of this paper. A case study of automotive engineering is vastly highlighted in this treatise.[11]

Mankind will witness new scientific prowess in this century. Science and technology are moving forward at a rapid pace. Environmental remediation will be the cornerstone of every scientific research today. In this paper, the authors deeply elucidate the scientific intricacies in the field of clean technology and its linkages with nanotechnology.[12–16] Green nanotechnology and green sustainability will also unravel the deep scientific truth and the scientific progress in science and technology globally today.

13.5.2   *ENVIRONMENTAL BIOTECHNOLOGY AND THE VAST VISION FOR THE FUTURE*

Environmental biotechnology is a revolutionary branch of science and technology today. Scientists and engineers are today in the forefront of research endeavor as regards application of environmental biotechnology to human society. Environmental biotechnology is biotechnology that is applied to and used to study the natural environment. The International Society of Environmental Biotechnology defines environmental biotechnology as "the development, use and regulation of biological systems for remediation of contaminated environment and for environment friendly processes."[27,28] The scientific scenario in the field of environmental remediation is extremely grave and thought-provoking. In this paper, the authors deeply reiterate the scientific success and the scientific ingenuity in the field of both environmental protection and nanotechnology.[17–20]

13.6   CLEAN NANOTECHNOLOGY MANUFACTURING

Clean nanotechnology manufacturing is the need of mankind and scientific advancements today. Technological challenges and the vast scientific and engineering profundity will surely unravel the hidden concepts of nanotechnology, engineered nanomaterials, and nanomaterials. Global research and development initiatives need to be streamlined and re-organized as regards environmental protection and application of nanotechnology. Nanomanufacturing is the need of science and technology today. Clean technology manufacturing and nanomanufacturing are the pillars of scientific and engineering endeavor today. Nanotechnology is creating wonders and is poised toward a new revolution. The world of manufacturing engineering and nanomanufacturing and its vast and versatile scientific ardor and vision will surely pave the way toward a newer era in nanotechnology and other branches of applied sciences. Civilization and its vast scientific grit and prowess are the pathways toward a newer scientific generation in the field of nanotechnology.[21–24] Science of clean technology manufacturing is today surpassing vast and versatile scientific boundaries. Manufacturing engineering today is in the midst of introspection and vision.

13.7 INVESTMENTS AND COMMERCIALIZATION OF CLEAN NANOTECHNOLOGY

Investments and commercialization of clean nanotechnology should be in war-footing as science and engineering moves forward. Human scientific endeavor in clean nanotechnology today is in the middle of deep scientific introspection and vast scientific grit and integrity. The commercialization of the technology of nanoscience and nano-engineering should be more streamlined and more re-organized as nanotechnology confronts some fundamental issues. Clean technology and green nanotechnology are the needs of civilization today. The active participation of civil society, governments, scientists, and engineers will surely pave the way toward a new epoch in the field of clean nanotechnology, green nanotechnology, and nanomanufacturing—and thus, the need of vast investments and commercialization of clean nanotechnology. The application of nanotechnology in environmental protection and the field of environmental nanotechnology are areas of vast scientific vision and scientific profundity today. In this review paper, the authors deeply reiterate the scientific success, the scientific integrity, and technological vision in the application of clean technology in the furtherance of science and engineering in the global scenario.[25–28]

13.8 A DEBATE: ADVANTAGES AND LIMITATIONS

A serious debate is in the forefront of scientific research pursuit—the application of nanotechnology in human society. The ethics of application of nanotechnology in human progress cannot be ignored as nanomaterials are used in different areas of science and engineering. Technology and engineering revamping are the necessities of the day as civilization is confronted with the monstrous issue of groundwater and drinking water heavy metal contamination. The debate of the application of nanotechnology in water and wastewater treatment is enormous and path-breaking. Heavy metal and arsenic drinking water contamination are in the middle of deep scientific introspection and scientific contemplation. Advantages of nanomaterials applications in diverse areas of science and engineering are enormous rather than limitations. Health and safety issues in nanomaterials and nanotechnology applications are veritably changing the path of science and engineering globally.

13.9   CONCLUSION AND SCIENTIFIC PERSPECTIVES

Nanotechnology and nano-engineering are the challenges and the vision of science and technology today. Research and development forays will surely open up new doors of scientific innovation in nanotechnology and nanomaterials applications. The challenges and the vision of scientific validation today are ever-growing and path-breaking. Clean technology concepts as regards nanotechnology applications needs to be vastly envisioned as science and engineering confronts with major issues of environmental engineering and chemical process engineering. The status of environmental engineering is grave and thought provoking today. Arsenic and heavy metal groundwater contamination are really confronting the scientific and engineering domain today. In such a crucial juncture of science and time, nanotechnology is the only answer to the many questions of environmental remediation. The concepts of unit of operations in chemical engineering or chemical engineering mass transfer operations are the needs of civilization and needs to be re-envisaged and reframed as civilization treads forward. In this entire treatise, the authors discuss with vision and insight the entire range of nanotechnology applications in environmental remediation. Adsorption, bioremediation, and removal of hazardous materials with the help of nanomaterials are the various avenues of scientific research pursuit today. The targets, the might, and the scientific divination of nanomaterials applications in human progress will surely open up new doors of scientific innovation in decades to come. In the similar vein, academic and scientific rigor in the field of nanomaterials and clean technology will surely strive forward in unfolding the scientific intricacies in the vast holistic domain of nanotechnology. Scientific and engineering perspectives in environmental remediation and clean nanotechnology will usher in a new era in the field of validation of science. This vision is elucidated in details in this chapter.

KEYWORDS

- **nanotechnology**
- **nanomaterials**
- **green**
- **vision**
- **sustainability**
- **environment**

IMPORTANT WEBSITES FOR REFERENCE

1. *https://www.understandingnano.com/cleaning.html*
2. *https://en.wikipedia.org/wiki/Green_nanotechnology*
3. *https://www.omicsonline.org/.../nanotechnology-as-a-tool-for-enhanced-renewable-en...*
4. *https://www.azonano.com/article.aspx?ArticleID=5017*
5. *https://www.forbes.com/.../nanotechnology-applications-that-can-change-the-world-al.*
6. *https://www.nanowerk.com/nanotechnology-in-green-industries.php*
7. *https://www.azocleantech.com/article.aspx?ArticleID=330*
8. *https://en.wikipedia.org/wiki/Green_nanotechnology*
9. *https://www.nanowerk.com/nanotechnology-in-green-industries.php*
10. *www.merid.org/en/Content/News.../Nanotechnology_and.../green_nanotech.aspx*
11. *https://www.biotecharticles.com/Nanotechnology.../Green-Nanotechnology-Its-Definit...*
12. *https://www.ncbi.nlm.nih.gov/pmc/articles/PMC4005277/*
13. *https://www.tandfonline.com/loi/ugnj20*
14. *https://www.mdpi.com/journal/materials/special_issues/green_nanotechnology*
15. *https://ostaustria.org/bridges.../2294-green-nanotechnology-a-path-to-sustainability*
16. *https://www.azocleantech.com/article.aspx?ArticleID=330*
17. *https://www.understandingnano.com/environmental-nanotechnology.html*
18. *https://www.journals.elsevier.com/environmental-nanotechnology-monitoring-and-ma...*
19. *https://www.nanowerk.com/nanotechnology-and-the-environment.php*

REFERENCES

1. *The Royal Society Report 2004; Nanoscience and Nanotechnologies: Opportunities and Uncertainties*; The Royal Society and The Royal Academy of Engineering: London, United Kingdom, 2004.
2. Neumayer, E. *United Nations Development Programme Report 2011; Sustainability and Inequality in Human Development*; Human Development Reports, Research Paper 2011/04; New York, November 2011.
3. Markulev, A; Long, A. *Australian Government Productivity Commission Report 2013; On Sustainability: An Economic Approach*; Staff Research Note: Canberra, Australia, May 2013.
4. Kates, R. W.; Parris, T. M.; Leiserowitz, A. A. What Is Sustainable Development? Goals, Indicators, Values and Practice, Environment. *Sci. Pol. Sustain. Develop.* **2005,** *47* (3), 8–21.
5. Kuhlman, T.; Farrington, J. What Is Sustainability? *Sustainability* **2010,** *2*, 3436–3448.
6. Werkneh, A. A.; Rene, E. R. Applications of Nanotechnology and Biotechnology for Sustainable Water and Wastewater Treatment. In *Water and Wastewater Treatment Technologies, Energy, Environment and Sustainability*; Bui, X.-T., et al., Eds.; 2019; Springer Nature: Singapore. https://doi.org/10.1007/978-981-13-3259-3_19, pp 405-430.
7. Kunduru, K. R.; Nazarkovsky, M.; Farah, S.; Pawar, R. P.; Basu, A.; Domb, A. J. Nanotechnology for Water Purification: Applications of Nanotechnology Methods in Wastewater Treatment. In *Water Purification*; Grumezescu, A., Ed.; Academic Press, Elsevier: Amsterdam, 2017; pp 33–74.
8. Das, R.; Vecitis, C. D.; Schulze, A.; Cao, B.; Ismail, A. F.; Lu, X.; Chen, J.; Ramakrishna, S. Recent Advances in Nanomaterials for Water Protection and Monitoring. *Chem. Soc. Rev.* **2017,** DOI: 10.1039/c6cs00921b.
9. Verma, A.; Sharma, M.; Tyagi, S. Green Nanotechnology, Research and Reviews. *J. Pharm. Pharma. Sci.* December **2017,** *5* (4), pp 60–66.
10. Nath, D.; Banerjee, P.; Das, B. "Green Nanomaterial"—How Green They Are as Biotherapeutic Tool. *J. Nanomed. Biotherapeu. Discov.* **2014,** *4* (2), 1–11.
11. Dhingra, R.; Naidu, S.; Upreti, G.; Sawhney, R. Sustainable Nanotechnology: Through Green Methods and Life-Cycle Thinking. *Sustainability* **2010,** *2*, 3323–3338. DOI: 10.3390/su2103323.
12. Palit, S. Microfiltration, Groundwater Remediation and Environmental Engineering Science—A Scientific Perspective and a Far-Reaching Review. *Nat. Environ. Poll. Technol.* **2015,** *14* (4), 817–825.
13. Palit, S.; Hussain, C. M. Biopolymers, Nanocomposites, and Environmental Protection: A Far-Reaching Review. In *Bio-Based Materials for Food Packaging*; Ahmed, S., Ed.; Springer Nature: Singapore Pte Ltd., 2018; pp 217–236.
14. Palit, S.; Hussain, C. M. Nanocomposites in Packaging: A Groundbreaking Review and a Vision for the Future. In *Bio-Based Materials for Food Packaging*. Ahmed, S., Ed.; Springer Nature: Singapore Pte Ltd., 2018; pp 287–303.
15. Palit, S. Advanced Environmental Engineering Separation Processes, Environmental Analysis and Application of Nanotechnology—A Far-Reaching Review. In *Advanced Environmental Analysis—Application of Nanomaterials*, Volume 1; Hussain, C. M.,

Kharisov, B., Eds.; The Royal Society of Chemistry: Cambridge, United Kingdom, 2017; pp 377–416.

16. Hussain, C. M.; Kharisov, B. *Advanced Environmental Analysis—Application of Nanomaterials*, Volume 1; The Royal Society of Chemistry: Cambridge, United Kingdom, 2017.

17. Hussain, C. M. Magnetic Nanomaterials for Environmental Analysis. In *Advanced Environmental Analysis—Application of Nanomaterials*, Volume 1; Hussain, C. M., Kharisov, B., Eds.; The Royal Society of Chemistry: Cambridge , United Kingdom, 2017; pp 3–13.

18. Hussain, C. M. *Handbook of Nanomaterials for Industrial Applications*. Elsevier: Amsterdam, Netherlands, 2018.

19. Palit, S.; Hussain, C. M. Environmental Management and Sustainable Development: A Vision for the Future. In *Handbook of Environmental Materials Management*; Chaudhery Mustansar Hussain, Ed.; Springer Nature: Switzerland A.G., 2018; pp 1–17.

20. Palit, S.; Hussain, C. M. Nanomembranes for Environment. In *Handbook of Environmental Materials Management*; Chaudhery Mustansar Hussain, Ed.; Springer Nature: Switzerland A.G., 2018; pp 1–24.

21. Palit, S.; Hussain, C. M. Remediation of Industrial and Automobile Exhausts for Environmental Management. In *Handbook of Environmental Materials Management*; Chaudhery Mustansar Hussain, Ed.; Springer Nature: Switzerland A.G., 2018; pp 1–17.

22. Palit, S.; Hussain, C. M. Sustainable Biomedical Waste Management. In *Handbook of Environmental Materials Management*; Chaudhery Mustansar Hussain, Ed.; Springer Nature: Switzerland A.G., 2018; pp 1–23.

23. Palit, S. Industrial vs Food Enzymes: Application and Future Prospects. In *Enzymes in Food Technology: Improvements and Innovations*; Kuddus, M., Ed.; Springer Nature: Singapore Pte. Ltd., Singapore, 2018; pp 319–345.

24. Palit, S.; Hussain, C. M. Green Sustainability, Nanotechnology and Advanced Materials—A Critical Overview and a Vision for the Future. In *Green and Sustainable Advanced Materials*, Volume 2, Applications; Ahmed, S., Chaudhery Mustansar Hussain, Eds.; Wiley Scrivener Publishing: Beverly, Massachusetts, USA, 2018; pp 1–18.

25. Palit, S. Recent Advances in Corrosion Science: A Critical Overview and a Deep Comprehension. In *Direct Synthesis of Metal Complexes*; Kharisov, B. I., Ed.; Elsevier: Amsterdam, Netherlands, 2018; pp 379–410.

26. Palit, S. Nanomaterials for Industrial Wastewater Treatment and Water Purification. In *Handbook of Ecomaterials*; Springer International Publishing: Switzerland A.G., 2017; pp 1–41.

27. www.wikipedia.com (accessed April 25, 2019).

28. www.google.com (accessed April 25, 2019).

CHAPTER 14

Wear Studies on Aluminum Metal Matrix Composite Prepared from Discarded Tea Extract and Coconut Shell Ash Particles

L. FRANCIS XAVIER*, K. SHIVA KUMAR, G. RAVICHANDRAN, N. SANTHOSH, and P. BENRAJESH

Assistant Professor, Department of Mechanical and Automobile Engineering, CHRIST (Deemed-to-be-University), Kanminike, Kumbalgodu, Mysore Road, Kengeri, Bangalore 560074, India

Corresponding author. E-mail: francis.xavier@chirstuniversity.in

ABSTRACT

In this present world, a lot of valuable natural resources are depleted for manufacturing products in order to meet the rising demand of the increasing population. As a result, in future, the industries may find it very difficult to manufacture the products due to the scarcity of raw materials. In this regard, researchers and industrialists around the globe have to join hands together to find an alternate method like recycling or utilizing low-cost reinforcement materials in preparing useful raw materials. In this work, aluminum metal matrixes composite was prepared by reinforcing discarded tea extract ash and coconut shell ash. Scanning electron microscope and EDAX test were used to characterize and study the morphology and chemical composition of the reinforcements used in this work for preparing the aluminum metal matrix composite. The mechanical properties of the prepared composite materials, like hardness and wear test, were conducted tested in order to investigate the influence of the reinforcement particles on the performance of the aluminum metal matrixes composite.

From the experimental results, it was observed that increasing the reinforcement content increases the hardness and also tends to improve the
resistance to wear rate.

14.1 INTRODUCTION

Composite materials are well-known for their favorable properties like low
density, high thermal stability, high electrical conductivity, higher stiffness
value, higher specific strength, adjustable coefficient of thermal expansion,
corrosion resistance, and good wear resistance, etc.[1-4] The performance
of the composite mainly depends on the following factors like (1) Nature
of interface between the matrix and the reinforcements, (2) shape of the
constituents, (3) size and distribution of the constituents, and (4) properties
of the matrix and reinforcement.

Generally, the composite material is composed of two phases namely the
base matrix phase and reinforcing phase. The base matrix phase is continuous
in nature, ductile, and less hard when compared with the reinforcement
phase. The main role of the matrix phase is to hold the reinforcements and
to share the applied load. On the other hand, the reinforcing phase is usually
stronger than the matrix phase and it is discontinuous in nature.

The composite materials can be broadly classified as metal matrix
composites (MMC), ceramic matrix composites (CMC), and polymer matrix
composites (PMC).[5] When compared with the other types of composites,
the MMCs possess significantly improved properties including high specific
strength; specific modulus, damping capacity, and good wear resistance
compared to unreinforced alloys. Thus they are used in various applications
like aerospace, automobile sectors, aircraft structures, sports, defense, etc.

Recently, due to an increase in the population and more demand
for products, a vast amount of research works are being carried out by
researchers around the globe. As a result, there is an increasing interest
among the researchers in preparing composite materials with low density
and low-cost reinforcements. As a result, there are many research works
conducted around the globe on the concept of low-cost reinforcement.[6-11]

In this work, Aluminum 6063 alloy was selected as the base matrix
material in preparing the composite material. Aluminum 6063 alloy is used
in various applications like heat sink sections, flexible assembly systems,
special machinery elements, railings, truck and trailer flooring, radiator,
and other heat exchanger applications.

14.2   DETAILS OF REINFORCEMENT MATERIALS

In this work, the discarded tea extract ash and coconut shell ash were used as the reinforcement materials in preparing the aluminum metal matrix composite. Initially, the reinforcement materials were dried in the sunlight for 5 days and then they were burnt in the furnace and the ash content was collected. Then, by using a sieve of mesh size 300, the ash particles were sieved and they were collected.

14.2.1   COCONUT SHELL ASH

Figure 14.1(a–d) shows the scanning electron microscope (SEM) image of the coconut shell ash (CSA) at various magnifications. Figure 14.1(e) shows the XRF analysis of coconut shell ash revealing the major elements present in it. Table 14.1 shows the weight percentage of the various elements present in CSA ash particles.

(a)   (b)

(c)   (d)

FIGURE 14.1   (a–d) SEM images of the CSA-ash particles at various magnifications.

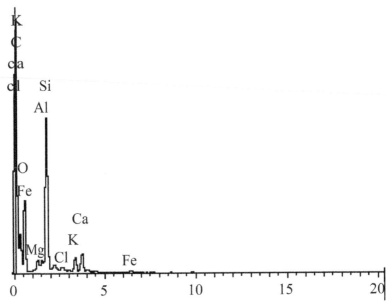

FIGURE 14.1   (e) Energy dispersive X-ray analysis (EDAX) test report of coconut shell ash particles.

TABLE 14.1   Details of the Various Elements Present in CSA Ash Particles.

Element	App conc.	Intensity Corrn.	Weight %	Weight % sigma	Atomic %
C K	32.80	0.5048	37.93	4.87	48.96
O K	39.69	0.5489	42.21	3.36	40.91
Mg K	1.28	0.7599	0.98	0.12	0.63
Al K	1.01	0.8499	0.69	0.10	0.40
Si K	20.23	0.9148	12.91	1.03	7.13
Cl K	0.59	0.7728	0.44	0.09	0.19
K K	2.82	1.0205	1.61	0.16	0.64
Ca K	3.88	0.9570	2.37	0.22	0.92
Fe K	1.16	0.7943	0.85	0.15	0.24
Total			100		

14.2.2   TEA EXTRACT

Figure 14.2(a–d) reveals the SEM image of the tea extract ash at various magnifications. Figure 14.2(e) shows the XRF analysis of tea extract ash revealing the major elements present in it. Table 14.2 shows the weight percentage of the various elements present in tea extract ash particles.

FIGURES 14.2 (a–d) SEM images of tea extract ash particles at various magnifications.

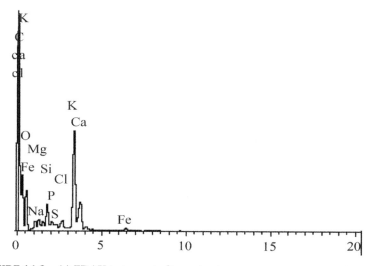

FIGURE 14.2 (e) EDAX test report of tea extract.

TABLE 14.2   Details of the Various Elements Present in Tea Extract Ash Particles.

Element	App conc.	Intensity corrn.	Weight %	Weight % sigma	Atomic %
C K	62.70	0.8215	41.75	3.02	55.04
O K	26.18	0.4112	34.83	1.93	34.47
Na K	1.73	0.7931	1.19	0.13	0.82
Mg K	1.62	0.7335	1.21	0.11	0.79
Si K	3.93	0.9106	2.36	0.16	1.33
P K	0.95	1.2915	0.40	0.09	0.20
S K	0.49	0.9582	0.28	0.07	0.14
Cl K	1.33	0.8403	0.87	0.09	0.39
K K	23.76	1.0656	12.20	0.66	4.94
Ca K	7.30	0.9337	4.28	0.27	1.69
Fe K	0.93	0.7932	0.64	0.13	0.18
Total			100		

14.3   PREPARATION AND MECHANICAL PROPERTIES OF THE COMPOSITE

The composites were prepared by a stir casting process. At the outset, the reinforcements were preheated at 300 degrees Celsius for 15 min in order to eliminate the moisture content from the reinforcements. Then the base matrix Al6063 alloy was melted in the furnace by raising the temperature to reach the melting point of the Al 6063 alloy. Then the molten melt was allowed to cool until it reaches semisolid state and the reinforcements were mixed and were stirred using a stirrer for 15–20 min. Then the temperature was increased to 750 degrees Celsius and the stirring was done for 15 more minutes and the molten melt was poured in to the die and was allowed to cool in the room temperature. Figure 14.3 and Figure 14.4 show the SEM image of the prepared sample-1 and sample-2. From the SEM image, we can see the uniform distribution of the reinforcement particles in the base matrix alloy.

The mechanical properties of the prepared samples were investigated; in this regard, the samples were subjected to a hardness test. Table 14.3 shows the results of the hardness value of the prepared samples. From the table, it is clear that the hardness value of the sample-2 is greater than that of sample-1. The presence of the hard reinforcement materials in the composites has played a vital role in increasing the hardness value of the composites. Thus, by increasing the reinforcement content, the hardness value of the composite tends to increase.

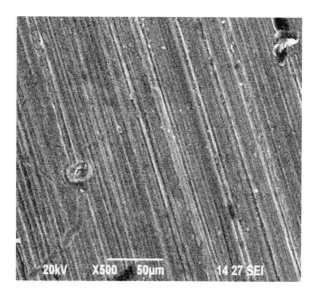

FIGURE 14.3 SEM Image of sample-1.

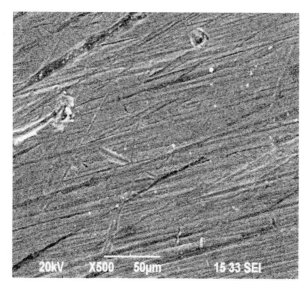

FIGURE 14.4 SEM Image of sample-2.

TABLE 14.3 Mechanical Properties of the Samples Taken for the Investigation.

Sl. No.	Material details	Hardness value (BHN)
1.	Sample 1: (Al6063 Alloy+10 wt % of Tea Extract +10 wt % of CSA)	65
2.	Sample 2: (Al6063 Alloy+15 wt % of Tea Extract +15 wt % of CSA)	88

14.4 WEAR TEST

Investigation on the Wear rate of the material is a very important criterion because it ensures the material's reliability in applications where they come in contact with other surfaces. In this work, the prepared composite materials were subjected to wear test. The wear test was conducted using pin-on-disc apparatus, Ducon make. The test was performed at 2 KgF load and 716 rpm for 6 min.

From the wear test result as shown in the Figures 14.5 and 14.6, we can see that for sample-1, the wear rate was 124 micrometers, whereas for the sample-2 the wear rate was only 70 micrometers. This indicates that the sample-1 has experienced more wear rate when compared with that of the sample-2. From Table 14.3, it is clear that the hardness of the sample-2 is 88 BHN whereas for sample-1 it is only 65 BHN. Thus, it is clear that increasing the reinforcement content increases the hardness of the composite and the wear rate. Further, on looking into Figures 14.5 and 14.6., the COF is 5.1N for sample-1 and for sample-2 it only 4.6N. As the COF value of sample-1 is higher, the sample-1 has experienced a higher amount of wear rate.

14.5 CONCLUSIONS

In this work, the discarded tea extract ash and coconut shell ash were used as the reinforcement materials in preparing AMMC. Stir casting method was used to prepare the composites. Further, from the chemical analysis of the reinforcement material, the presence of silica, carbon, iron, and magnesium makes them suitable materials in preparing the composite. From the experimental results, it is clear that sample-2 has better resistance to wear rate when compared with the sample-1 which indicates that increasing the reinforcement content increases the hardness value and resistance to the wear rate of the composite.

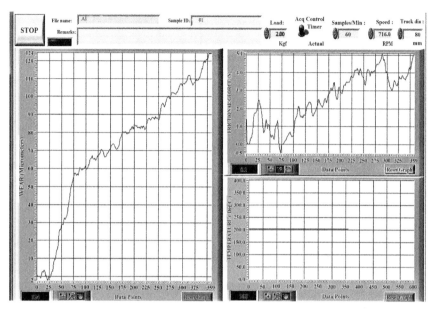

FIGURE 14.5 Wear test results of sample-1.

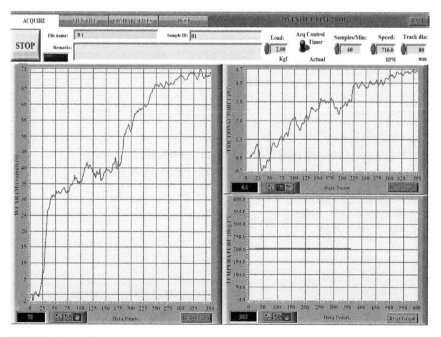

FIGURE 14.6 Wear test results of sample-2.

KEYWORDS

- **tea extract**
- **coconut shell ash**
- **discarded materials**
- **EDAX test**
- **aluminum metal matrix composite**

REFERENCES

1. Akbulut, M.; Yilmaz, D. F. Dry Wear and Friction Properties of δ-Al2O3short Fiber Reinforced Al-Si (LM 13) Alloy Metal Matrix Composites. *Wear* **1998**, *215* (1–2), 170–179.
2. Skolianos, S.; Kattamis, T. Z. Tribological Properties of SiCp-Reinforced Al-4.5% Cu1.5% Mg Alloy Composites. *Mater. Sci. Eng.* **1993**, *A163*, 107–115.
3. Surappa, M. K.; Prasad, S. V.; Rohatgi, P. K. Wear and Abrasion of Cast Al-alumina Particle Composites. *Wear* **1982**, *77* (3), 295–302.
4. Cao, L.; Wang, Y.; Yao, C. K. The Wear Properties of an SiC–Whisker Reinforced Aluminium Composite. *Wear* **1990**, *140*, 273–277.
5. Feest, E. Interfacial Phenomena in Metal-Matrix Composite. *Composites* **1994**, *25* (2), 75–86.
6. Lancaster, L.; Lung, M. H.; Sujan, D. Utilization of Agro-Industrial Waste in Metal Matrix Composites: Towards Sustainability. *Int. J. Envir. Ecol. Eng.* **2013**, *7* (1), 35–43.
7. Kamble, A.; Kulkarni, S. G. Microstructural Examination of Bagasse Ash Reinforced Waste Aluminium Alloy Matrix Composite. *AIP Conf. Proc.* **2019**, *2105* (020011), 1–7. https://doi.org/10.1063/1.5100696.
8. Matsunaga, T.; Kim, J. K.; Hardcastle, S.; Rohatgi, P. K. Crystallinity and Selected Properties of Fly Ash Particles. *Mat. Sci. Eng.* **2002**, *A325*, 333–343.
9. Koczak, M. J.; Prem Kumar, M. K. Emerging Technologies for the In-situ Production of MMCs. *JOM* **1993**, *45* (1), 44–48.
10. Maity, P. C.; Chakraborty, P. N.; Panigrahi, S. C. Preparation of Aluminium-Alumina In-Situ Particle Composite by Addition of Titania to Aluminium Melt. *Scripta. Metall. Mater.* **1993**, *28* (5), 549–552.
11. Kumar, B. P. Microstructure and Mechanical Properties of Aluminium Metal Matrix Composites With Addition of BLA by Stir Casting Method. *Tran. Nonferr. Metal Soc. China* **2017**, *27* (12), 2555–2572.

CHAPTER 15

Study on Fretting Wear Characteristics of Aluminum-Based Metal Matrix Hybrid Composites Fabricated by Friction Stir Welding Technique

M. SADASHIVA[1*], M. R. SRINIVASA[1], N. SANTHOSH[2], G. RAVICHANDRAN[2], and V. SHARANRAJ[3]

[1]*Department of Mechanical Engineering, P.E.S.C.E, Karnataka, India*

[2]*Department of Mechanical and Automobile Engineering, Christ (Deemed to be University), Bangalore, Karnataka, India*

[3]*Department of Mechanical Engineering, S.J. Polytechnic, Bangalore, Karnataka, India*

*Corresponding author. E-mail: sadashiva015@gmail.com

ABSTRACT

Aluminum and its alloys, due to their excellent properties, have proved their application in several engineering domains. Among metal matrix composites (MMCs), aluminum and its alloys have emerged with special attention. The aluminum metal matrix has variety of applications with desired combinations due to the inclusion of copper, zinc, and magnesium elements in it. MMCs with different aluminum alloy combinations have a pivotal role in the fields of aviation and automobile domain due to their high strength to weight ratio and better fuel economy. For achieving proficient and desired mechanical properties like better corrosion resistance and wear, low coefficient of expansion, toughness, machinability and ductility, aluminum alloys are mainly used alongside a number of reinforcements. In this current work, AA 2024 was contrasted with 25-micron greenish silicon carbide ceramic particles and chopped E-glass reinforced fibers producing hybrid composites by stir casting followed by joining, carried with friction stir welding with low welding parameters. Research work focuses on evaluation of fretting wear characteristic of welded base and composites plates with linear reciprocating tribometer with variable load condition due to ceramic particulates in composite and nugget of weld part exhibits lower wear rate and result was tabulated.

15.1   INTRODUCTION

The welding of aluminum and its alloys have always represented a great challenge for designers and technologies. One of the main limitations for the industrial application of these alloys is the difficulty in using conventional welding methods for joining. The purpose of this literature review is to obtain a comprehensive understanding on the subject related to this work. Friction stir welding is best suited for joining of aluminum-based composites. Particulate-reinforced and chopped fibers incorporated with metal matrix have gained attention and interests among materials scientists and engineers in recent years due to its excellent strength, high hardness, corrosion resistance, stiffness, wear resistive, and thermal properties.

Aluminum and its alloys have been studied extensively in recent times. Formation of metal matrix composites of different metals has shown a considerable improvement in their properties. In this research, the alloys that have been chosen are Al-2024. Aluminum has long been in use due to its excellent malleability, and this project work aims to investigate the extent to which it can be used in fields that require high strength and hardness, which is impossible to achieve in pure aluminum. Within the different alloys chosen, there is a difference made in terms of the composition of the different reinforcements used. The reinforcements chosen for the present study are silicon carbide (SiC) and E-glass fiber, with the former used in a much larger quantity than the latter. In the present study, the tensile strength of the metal matrix composites (MMCs) obtained as the result is investigated and the results are then compared in order to achieve further inferences. Shivanand et al.[1] carried out research on the effect of E-glass fiber reinforcement in composite materials with different percentage of fiber on impact test with low impact velocity. Results exhibits clearly that the percentage of e-glass fiber in composites increases its toughness and strength considerably. Yahya et al.[3] analyzed that Weldability A2124 containing 25% SiC particles with T4 heat-treated aluminum MMC plates was investigated at low welding parameter by FSW process. Microstructure, ultimate tensile strength, surface roughness and micro hardness determination have been carried out to evaluate the weld zone characteristics of friction stir welded MMC plates. The fracture surface resulting from the tensile test revealed a mixed brittle-ductile fracture mode and also showed the superiority of bonding between the SiC particles and Al matrix. Mishra et al. [2] reported that FSW, as latest and advanced promising solid-state joining method, could be best for light metals and composites, like aluminum welding, and reported about temperature distribution during welding and effect of heat treatment on weld quality. Ramesha et al.[4] clearly explained the method of joining and focused the effect of tool materials on welding. Santhosh et al.[5] focussed on the thermomechanical characterization of weld zones, advancing side and retreating side were clearly evaluated in their work. Further, Ravichandran et al.[6,7] have carried out their research on the effect of heat treated HNT reinforcements on the physic-mechanical properties of composites which have given information on the testing techniques and methodologies to be followed for efficient characterization. Ramesha et al.[8] have carried out extensive research on optimization of process parameters for machining of "glass fiber reinforced polymer," (GFRP) composites by abrasive water jet machining that effectively gives an overview of the design of experiments for the research work postulated. Further, Santhosh et al.[9] have carried out research on vibration characteristics of aluminum 5083/ SiCp/fly ash composites which explains the importance of characterization of aluminum composites for real time applications.

The research work formulated have objectives, which are as follows:

1) Selection of proper matrix material and reinforcements in order to produce high standard MMC having superior properties and behavior.
2) To fabricate high quality MMC material through stir casting.
3) Weld the composite plates so as to obtain sound weld joint using friction stir welding (FSW).
4) To create standard fretting wear test specimens conforming to ASTM standards using wire electric discharge machining (EDM).
5) Conduct fretting wear test using linear reciprocating tribometer (LRT).

15.2 MATERIALS AND METHODS

15.2.1 SELECTION OF MATRIX AND REINFORCEMENTS

One of the most important series of aluminum alloys AA 2024 was selected for the investigation. It has greater acceptance in automobile, aerospace, and industrial applications. The chemical composition of AA 2024 is given in Table 15.1.

TABLE 15.1 Chemical Composition of AA 2024.[10]

Si	Fe	Cu	Mn	Mg	Cr	Zn	Ti	Remainder
0.5	0.5	3.8–4.9	0.3-0.9	1.2–1.8	0.1	0.25	0.15	0.15

Silicon carbide is commonly used reinforcement with aluminum because SiCp is easily available and has good wettability with aluminum alloys. Density of SiC is 3.2 g/cm^3. It is more preferred as it is a good reinforcement for aluminum-based MMC, that improves its strength and hardness. Figures 15.1–15.3 give the photographic representation of the materials used for the fabrication of composite plates.

FIGURE 15.1   Al 2024 Ingot.

Due to light weight and high strength characteristics, glass fibers are most eminent option for production of hybrid composites. Among glass fibers, E-glass fibers are economical and low density. Due to high chemical resistance, electrically insulated, and high stiffness, in most of the aerospace and marine applications E-glass dominantly takes major part. Chemical composition of E-glass fiber is given in Table 15.2.

FIGURE 15.2   SiCp.

FIGURE 15.3   E-glass fiber.

TABLE 15.2   Chemical Composition of E-glass Fiber.[11,12]

SiO$_2$	Al$_2$O$_3$	CaO	MgO	B$_2$O$_3$
54.3%	15.2%	17.2%	0.6%	8.0%

15.2.2 STIR CASTING FOR PREPARATION OF AMMC

The aluminum alloy 2024 was loaded and melted in an open coke graphite coated crucible furnace by heating at 800°C. The wt% 3 and 7, shortly chopped E-glass fibers (3 mm length, 12 μm diameter) and fine greenish SiCp powder, is preheated at 600 and 900°C, respectively, before adding in molten alloy. At first, preheated E-glass was added in the liquid melt and the slurry was continuously stirred using four-bladed stirrer. Then, the preheated SiC is added into the liquid melt with continued stirring. The stirring of mixture was carried out for 4–5 min (stir period) to obtain homogeneity of mixture. The stirring speed of 700 RPM was maintained throughout the process. Then, the temperature of the mixture was checked using pyrometer and when the mixture attained 700°C, it was poured in the mould that was prepared by green sand molding as shown in Figures 15.4–15.6. The liquid composite was allowed to solidify to obtain aluminum composite plates of dimension (50 × 50) mm.

FIGURE 15.4   Open hearth furnace.

15.2.3 FRICTION STIR WELDING

The friction stir welding (FSW) was first developed by Thomas at TWI (the welding institute). FSW is used to produce low cost and high performance joints. FSW method can be used to avoid structural defects such as porosity, fume, shrinkage, etc. It is ideal for aluminum composites. FSW is performed to form butt joint between two composite plates that were casted earlier. Here, welding parameters such as tool rotating speed and feed were varied and number of specimens were obtained.

FIGURE 15.5 Moulds.

FIGURE 15.6 Cast plate.

The obtained AMC plates were joined by welding process. FSW was operated to form efficient butt joint between two plates. The machine used for the FSW is shown in Fig. 15.7. During the welding operation, welding parameters such as tool rotational speed and feed were considered for the analysis. The welding speed operated was 600 RPM. The welding feed operated was 40 mm/min. Single pass butt weld was made with fixed tilt angle. The tool used in the welding is shown in Fig. 15.9. The tool consists of 50 mm shoulder and 5 mm cylindrical pin at one end. The efficient joint was obtained to produce plate of dimension (100 × 100) mm as shown in Fig. 15.8. Figures 15.7–15.9 give the photographic views of the FSW machine used in the process, welded plate and the tool used for the process respectively.

FIGURE 15.7   FSW machine.

15.2.4   *MACHINING OF WELDED PLATES TO STANDARD SPECIFICATIONS*

The welded plates were then cut into the required sizes using wire EDM as per the ASTM standards for fretting test (length 40 mm × width 40 mm × thickness 5 mm).

15.3   EXPERIMENTATION

The experimentation was carried out in the research in accordance with the process parameters given in Tables 15.3 and 15.4 respectively, representing two cases of research hypothesis postulated.

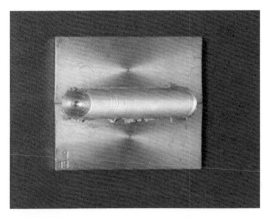

FIGURE 15.8 Welded plate.

FIGURE 15.9 Tool.

Case 1. Process parameters

TABLE 15.3 (AA2024 + 0% E-Glass fiber + 0% SiC)

Spindle speed	600 RPM
Traverse speed	40 mm/min
Plunger depth	5.6 mm
Pre heating time	10 s
Tool tilt angle	0

Case 2. Process parameters

TABLE 15.4   (AA2024 + wt% 3E-Glass fiber + wt% 7SiC)

Spindle speed	600 RPM
Traverse speed	40 mm/min
Plunger depth	5.6 mm
Pre heating time	10 s
Tool tilt angle	0

15.3.1   *FRETTING WEAR TEST*

Tomlinson termed the fretting wear in 1927. Fretting wear is also known as vibrational wear, chafing, fatigue, wear oxidation, fretting fatigue, and corrosion. Fretting wear is caused by the small amplitude movement (1–300 μm) with high frequency oscillating movements. When small amplitude reciprocating motion or sliding takes place between the contacting surfaces for a large number of cycles, fretting occurs (refer to Figures 15.10–15.12 for photographic representation of the fretting wear test process).

FIGURE 15.10   Fretting wear testing.

Fretting wear occurs in all vibrating machines, joints like bolts, press-fitted, pins, rivets, and keys. It also occurs in the components that are constrained, oscillating splines, couplings, bearings, clutches, spindles and seals, universal joints, and shackles where relative sliding in micron level is allowed. Even though the amplitude of reciprocating movements is as low as 0.1 micron, the failure of components occurs if sliding is continued for one million cycles or more. The fatigue cracks are initiated by fretting that will result in fatigue failure in shafts and other load carrying components. The factors like displacement amplitude, normal load, properties of material, frequency of oscillations, and humidity will greatly affect the wear characteristics. Fretting wear can be affected by temperature. The co-efficient of friction of surfaces depends on the amplitude of the relative oscillation. Table 15.5 gives the specification of the LRT equipment used in the research.

FIGURE 15.11   Fretting specimen (plate).

FIGURE 15.12   Pin.

TABLE 15.5   Specification of the Linear Reciprocating Tribometer (LRT).

S. No.	Specification	Values
1	Cylindrical pin indenter	8 mm diameter × 15 mm long
2	Spherical ball indenter	8 mm diameter
3	Rectangular flat pin	15 × 10 × 6 mm
4	Flat specimen	40 × 40 × 5 mm
5	Normal load	20–200 N
6	Frequency	5–50 Hz
7	Stroke length	1–20 mm
8	Frictional force	Max 200 N
9	Temperature	Up to 300°C
10	Equipment size	600 × 520 × 920 mm

15.3.1.1   EXPERIMENTAL PROCEDURE FOR FRETTING WEAR TEST

The AA2024 plate and pin specimens were tested for their fretting wear characteristics. The plates are placed in the slot available in the LRT machine and tightened up firmly with the screws. The pin is fixed in the tool holder firmly with the help of Allen key drive and the surface of pin is brought in contact with the surface of the plate. A required load is applied on the holder by means of metallic discs. All the input data are given through computer system using 'WINDUCOM' software. Once all the setup is done, the experiment is carried out for a desired number of cycles of oscillations. Then, we obtain the output data directly from the WINDUCOM software. These results are acquired and reported.

15.3.1.2   INPUT PARAMETERS

TABLE 15.6   Input Parameters of the Fretting Wear Test.

Input parameters	Values
Stroke length	10 mm
Frequency	20 Hz
Time period	8.5 minutes
Applied load	20, 40, 60 N

15.3.1.3 OUTPUT PARAMETERS

TABLE 15.7 Output Parameters of the Fretting Wear Test.

Output parameters
Coefficient of friction
Frictional force
Weight loss

15.3.1.4 NUMBER OF TEST TRIALS

During the experimentation of specimens, three test trials were carried out on each type of series (S1, S2, S3, S4, S5, S6) of AL2024 by varying applied load in terms of 20, 40, and 60 N. The output results obtained for different series are tabulated, analyzed, and brought in the form of graphs. Then, the results and graphs are compared to find the wear characteristics of the base and composite AL2024 metal plates and the conclusions are drawn.

15.4 RESULTS AND DISCUSSIONS

TABLE 15.8 Comparative Results of Fretting Test for AA2024 Base Metal.

S. No.	Load (N)	Total initial weight (g)	Total final weight (g)	Reduction in weight (g)	% of reduction	Co-efficient of friction (μ)	Frictional force (N)
S1, S2, S3 (20S1F1) AA2024 base plate welded by FSW with the tool rotational speed of 600 RPM and feed rate of 40 mm/min for varying applied load in terms of 20, 40, and 60 N.							
1	20	23.0474	22.9575	0.0899	0.390	0.2642	5.2866
2	40	22.9575	22.8283	0.1292	0.563	0.184	7.3381
3	60	22.6198	22.406	0.2138	0.945	0.162	11.048

Graphical representation of fretting wear results obtained for different specimens.

The comparison of coefficient of friction on these specimens at load 20, 40, and 60 N is depicted in the graph using colors of blue, red, and green, respectively.

TABLE 15.9   Comparative Results of Fretting Test for AA2024 + wt 3% E glass fiber + wt 7% SiC Composite Material.

S. No.	Load (N)	Total initial weight (g)	Total final weight (g)	Reduction in weight (g)	% of reduction	Co-efficient of friction (μ)	Frictional force (N)
S4, S5, S6 (2S1F1) AA2024 composite plate welded by FSW with the tool rotational speed of 600 RPM and feed rate of 40 mm/min for varying applied load in terms of 20, 40, and 60 N.							
1	20	21.1542	21.0944	0.0598	0.283	0.134	2.583
2	40	21.0944	21.0129	0.0815	0.386	0.1018	4.071
3	60	19.5317	19.3909	0.1408	0.721	0.1002	6.457

Graphs 1 and 2 show that relation between co- efficient of friction with time, progressively coefficient of friction decreases with time during fretting. Following graphs explain variations and effects of applied load, co-efficient of friction, weight loss on AA2024 composite samples welded at tool rotational speed 600 RPM and tool feed rate of 40 mm/min.

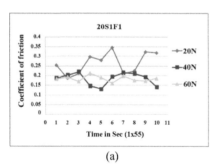
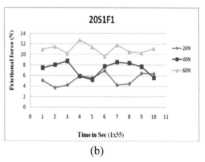

(a)   (b)

GRAPHS 1   (a and b)   Show the plot of coefficient of friction (COF), frictional force (FF) vs time (s) for 20S1F1.

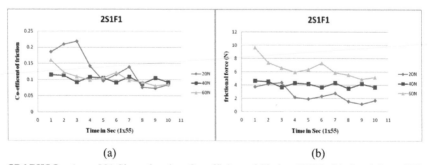

(a)   (b)

GRAPHS 2   (a and b) Show the plot of coefficient of friction (COF), frictional force (FF) vs time (s) for 2S1F1.

Graphs 3 and 4 clearly shows that increasing applied load on work during fretting progressively increases the wear rate. Due to sliding force and speed coefficient of friction decreases, because of the thermal softening effect between pin and work which smash out the oxide layer. The addition of SiC and E-glass fiber reinforcement to AA2024 matrix enhances the effective bonding, by allowing the large interfacial area of contact, which results in wear resistance.

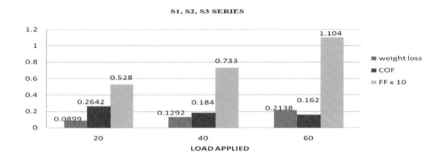

GRAPH 3 Graph shows the plot of weight loss, coefficient of friction (COF) and frictional force (FF) vs load applied for 20S1F1 (S1, S2, S3).

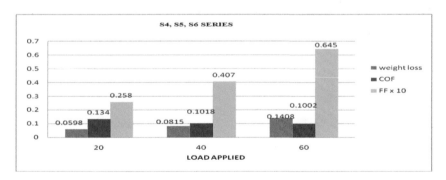

GRAPH 4 Graph shows the plot of weight loss, coefficient of friction (COF) and frictional force (FF) vs load applied for 2S1F1 (S4, S5, S6).

Graphs 5–7 reveal the comparative results of wear behavior between the base and composite specimens weld with speed of 600 RPM and feed rate of 40 mm/min for 20, 40, and 60 N.

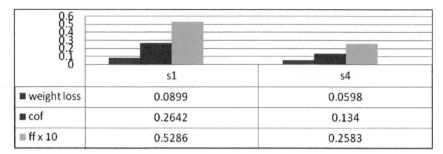

	s1	s4
■ weight loss	0.0899	0.0598
■ cof	0.2642	0.134
▦ ff x 10	0.5286	0.2583

GRAPH 5   The comparative results of weight loss, coefficient of friction (COF) and frictional force (FF) of the specimens S1 and S4 at 20 N.

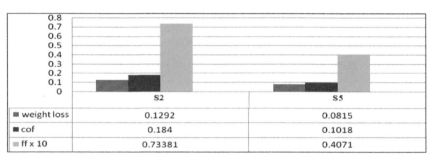

	S2	S5
■ weight loss	0.1292	0.0815
■ cof	0.184	0.1018
▦ ff x 10	0.73381	0.4071

GRAPH 6   The comparative results of weight loss, coefficient of friction (COF) and frictional force (FF) vs the specimens S2 and S5 at 40 N.

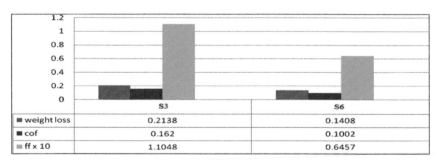

	S3	S6
■ weight loss	0.2138	0.1408
■ cof	0.162	0.1002
▦ ff x 10	1.1048	0.6457

GRAPH 7   The comparative results of weight loss, coefficient of friction (COF) and frictional force (FF) of the specimens S3 and S6 at 60 N.

The addition of SiC and E-glass fiber reinforcement to AA2024 matrix enhances the effective bonding, by allowing the large interfacial area of contact, which results in lesser wear resistance. The fretting wear rate decreases with the addition of silicon carbide and E-glass fiber reinforcement. It is herewith validated from the experimentations that the addition of the reinforcements to the matrix aluminum 2024 alloy increases its characteristics and thereby reduces its wear due to the fact that the heating of the specimens due to lower feed rates causes zones of controlled deformations and molten bands that will ultimately enhance its characteristic wear resistance properties.

15.5   SCANNING ELECTRON MICROSCOPE (SEM)

Scanning electron microscope (SEM) equipment was utilized to view the microstructure of the AA2024 alloy and SiC, E-glass fiber reinforced AA2024 composite samples.

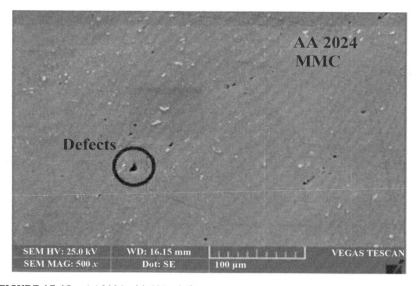

FIGURE 15.13   AA2024 with 0% reinforcement

The scanning electron microscope revealed the microstructure of the received samples. Fig. 15.13 shows the microstructure of the AA2024 sample with 0% reinforcement. Fig. 15.14 shows the microstructure of the AA2024 sample with 10% SiC reinforcement assured the excellent distribution of SiC reinforcement in the matrix phase.

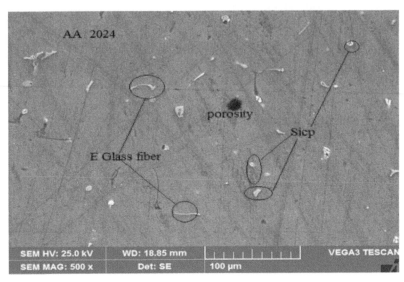

FIGURE 15.14   AA2024 composites.

15.6   CONCLUSIONS

1. The comparative results show that fretting wear resistance increase with addition of silicon carbide and E-glass fibers under the load condition. Wear resistance increased both due to the occurrence of SiC particles and E-glass fiber due to oxides build up on the sliding surface.
2. The inferences depicted dry sliding characteristics wear in AA 2024 base and AA2024 + wt 3% E-glass fiber + wt 7% SiC composite samples.
3. The result clearly shows that increasing the applied load on work during fretting progressively increases the wear rate. Further, due to the sliding force and speed, the coefficient of friction decreases because of the thermal softening effect between pin and work that removes the oxide layer. It is observed that wear rate increases with respect to increasing fretting load on tool and the wear was found to be maximum at 60 N and minimum at 20 N.
4. The fretting test on welded samples exhibits higher resistance to wear due to compaction of atoms at weld zones, precipitation of grains that leads to a resistant barrier for sling force leading to

subdued wear. The fretting wear rate (weight loss) is slightly more in welded base material compared to composites. About 18–20% reduction in wear rate found in welded AA2024-based composites.

KEYWORDS

- **metal matrix composites**
- **silicon carbide**
- **friction stir welding**
- **fretting wear**

REFERENCES

1. Prabhakar, K.; Shivanand, H. K. Experimental Studies on Mechanical Properties of E Glass Short Fibers & Fly Ash Reinforced Al 7075 Hybrid Metal Matrix Composites. *IJARME* **2012**, *2* (2), 2231–5950.
2. Yahya, B. Weldability of Metal Matrix Composite Plates By Friction Stir Welding At Low Welding Parameters. *Materiali in tehnologije/Mater. Technol.* **2011**, *455*, 407–412.
3. Mishra, R. S.; Ma, Z. Y. Friction Stir Welding and Processing. *Mater. Sci. Eng.: R: Rev. Report* **2005**, *50*, 1–78.
4. Ramesha, K; Santhosh, N; Sudersanan, P. D.; Bedi, R. A Comprehensive Review of Novelty of Friction Stir Welding of Aluminium-Magnesium Alloys for Advanced Engineering Applications, *J. Thin Films, Coating Sci. Technol. Appl.* **2018**, *5*(2), 7–14.
5. Santhosh, N; Mudabir, M. Thermomechanical Modeling and Experimental Evaluation of Friction Stir Welds of Aluminium AA 6061 Alloy, *IJERT* **2013**, *2*(8), 1494–1499, ISSN: 2278-0181.
6. Ravichandran, G.; Rathnakar, G.; Santhosh, N. Effect of Heat Treated HNT on Physico-mechanical Properties of Epoxy Nanocomposites. *Compos. Commun.* **2019**, *13*, 42–46.
7. Ravichandran, G.; Rathnakar, G.; Santhosh, N.; Chennakeshava, R.; Hashmi, M. A. Enhancement of Mechanical Properties of Epoxy/halloysite Nanotube (HNT) Nano-composites. *SN Appl. Sci.,* **2019**, *1*, 296. https://doi.org/10.1007/s42452-019-0323-9.
8. Ramesha, K; Santhosh, N; Kiran, K; Manjunath, N; Naresh, H. Effect of the Process Parameters on Machining of GFRP Composites for Different Conditions of Abrasive Water Suspension Jet Machining, *Arab. J. Sci. Eng.* **2019**, *44* (9), 7933–7943.

9. Santhosh, N; Kempaiah, U. N.; Venkateswaran, S. Vibration Mechanics of Hybrid Al 5083/SiCp/Fly Ash Composite Plates for its Use in Dynamic Structures, *J. Exp. Appl. Mech.* **2019**, *8*(1), 11–18.

10. Sadashiva, M.; Shivanand, H. K.; Vidyasagar, H. N. Characteristic Evaluation of Process Parameters of Friction Stir Welding of Aluminium 2024 Hybrid Composites. *Advances in Mechanical Design, Materials and Manufacture, American Institute of Physics (AIP) Conference Proceedings*, 2018, *1943*, 020054-1–020054-10; ISBN:9780735416383,https: //doi.org/ 10.1063/ 1.5029630.

11. Sharanraj, V.; Ramesha, C. M.; Kumar, V.; Sadashiva, M. Finite Element Analysis of Zirconia Ceramic Biomaterials Used in Medical Dental Implants", *SPRINGER-INTERCERAM, International Ceramic Review*, **2019**, *68*(3), 24–31, ISSN: 0020-5214.

12. Sadashiva, M; Srinivasa, M. R.; Sharanraj, V.; Santhosh, N. Experimental Investigation on Tensile Characteristics of Hybrid Nanocomposites Joints by FSW with Optimized Welding Parameters. *J. Nano Sci. Nano Eng. Appl.* **2019**, *9*(2), 1–13, ISSN 2231-1777.

Index

Printed and bound by CPI Group (UK) Ltd, Croydon, CR0 4YY

23/10/2024

01777702-0006